Explorations in Early Chinese Cosmology

Edited by
Henry Rosemont, Jr.

Papers presented at the Workshop
on Classical Chinese Thought
held at Harvard University
August 1976

Journal of the American Academy of Religion Studies

Volume L, Number 2

Scholars Press
Chico, California

☐ The Cover ──────────

The intersecting lines in the graphic sign represent the action of religious reality (the vertical line) upon the world of ideas (the horizontal line) by going back to the origin (the dot) from which reality and ideas come. In a general way, this is the intention of the Journal's thematic series in religious studies.

ISSN 0735–6919
ISBN 0–89130–656–0

Printed in the United States of America
on acid-free paper

The contributors and editor wish
to dedicate this volume to the
memory of their colleague

VITALY A. RUBIN

with respect for his scholarship
and with admiration for his spirit.

JAAR Thematic Studies L/2

Contents

Introduction

Henry Rosemont, Jr.

I

The impulse to reflect upon and attempt to comprehend the more general features of the universe is as widespread as self-consciousness; virtually every culture, past and present, has a view of the cosmos to be found in it.

The many and varied manifestations of this apparently universal human phenomenon should consequently be of interest to everyone who has similarly reflected, and attempted to comprehend. In most cultures, however, cosmological views were never set down explicitly, and must therefore be teased out of reconstructed myths, legends, art, artifacts, rituals, and other societal products. The teasing has often been brilliant, but is always complicated by the fact that these cultural objects seldom yield a unique interpretation—as the many heated disputes among scholars who have studied the same culture demonstrate./1/ An added difficulty is that the farther the culture under investigation is from Graeco-Roman culture, the less useful does pure and simple logic seem to become as an overall guiding principle of analysis and explication; one of the more unusual characteristics of ancient Egyptian divinities is that their worshippers seem to have simultaneously predicated incompatible properties of them.

Thus those cultures which have produced coherent cosmological writings deserve particularly close scrutiny. To be sure, texts, especially ancient texts, seldom yield unique interpretations either; philological and philosophical swords are drawn to this day about how to read even writings from our own cultural past, e.g., the *Timaeus*./2/ But texts do constrain interpretations much more narrowly than other cultural products. The archaic Chinese script differs greatly from modern English, yet it is modern English words and not Rorschach blots that the logographs must resemble. (And proffered textual interpretations are further constrained by the other cultural objects: given, for example, a specific textual account of life after death, we should expect funerary artifacts and burial practices to reflect that account.)

China is one of the relatively few cultures of antiquity to have produced texts containing explicit cosmological statements, and the papers gathered together in this volume explicate and analyze a number of those writings from over two millenia ago. First presented at the ACLS-Harvard Workshop on Classical Chinese Thought in 1976, these papers are remarkable in a number of ways.

First, as the date of their initial reading shows, these papers are now over seven years old, yet at the time they were finally prepared for the press by the editor, the authors did not feel the need to make *substantive* revisions of their contributions. That is to say, the continuing researches of the scholars represented herein, and the most recent scholarship of others, have suggested a new turn of phrase here, the revision of a footnote there, while the conclusions of the work(s), and the evidence and arguments on which the conclusions are grounded, are holding up well under the test of time.

Moreover, readers of this volume will, I believe, be surprised by how well integrated these essays are both with respect to methodology and outcome, even though at the time of original composition none of the seven authors were aware of what the others were doing; from Berkeley to Seattle to Moscow the scholarship was undertaken independently.

Even more surprising is that outcome: although the Chinese texts analyzed below (individually, for the most part) span the lengthy period from mid-Shang to early Han (roughly about 1350 to 150 B.C.E.), they collectively suggest that there are invariant themes in early Chinese cosmology, themes which survive the manifold disputations of the "hundred schools" of late Zhou Dynasty philosophy. It is a commonplace—perhaps too common—that the major philosophers of this period were social and moral thinkers, they were not metaphysicians. To the extent this generalization is warranted, the following essays suggest why: a very similar view of the cosmos was shared by all of them, with their disagreements centered on how best to order our individual and societal lives in accordance with the cosmological order. Their universe "spoke" to them—a metaphor to which I will return below—but like the oracle at Delphi, its pronouncements were open to interpretation and, consequently, philosophical disputation.

I do not need to describe that Chinese cosmological perspective further in this Introduction, nor to provide abstracts for the seven articles which follow. Basing themselves on close textual analyses and using historical and archaeological materials mixed with generous but measured dashes of speculation, the contributors to this volume individually and collectively provide us with a singular, convergent, and comprehensive early Chinese cosmology, rendering additional description on my part superfluous at best, misleading at worst; these papers have integrity on their own, and together.

Thus freed of narrative responsibilities I will follow the philosopher's bent of moving on to consider briefly some of the deeper implications of these studies of early Chinese cosmology. In my opinion this book is a ticket, first class, for a long-distance intellectual journey. The China of over two thousand years ago is a major stop along the way, but read carefully and reflectively the final destination will be at the frontiers of the reader's imagination. If the balance of this Introduction is not a useful sketch of that journey, the other contributors are not to blame, nor are they to be held responsible for the use (or misuse) to which I have put their work. Scholarly bouquets, in other words, should be thrown to the authors, brickbats to the editor.

II

Up to this point I have glossed over the question of *why* we should be interested in the cosmologies of other cultures, especially ancient cultures different from our own. (Where "our own" is to be described using terms such as "pulsar" and the like.) There are, of course, a number of straightforward answers to this question: such views will be of general antiquarian interest, and in the present case, of sinological interest more specifically. And if the impulse to wonder about the order of the cosmos is indeed universal, particular views should be of anthropological interest as well.

Moreover, new research on cosmological views may contribute to the manifold forms of the "one versus the many" debate among scholars. In the field of history, for example, "diffusionists" and "independent parallel developers" have been arguing for quite some time over whether there was a *locus classicus*—usually the Fertile Crescent ca. 2500 B.C.E., occasionally pushed further back—from which all other cultures draw, or whether some cultures, at least, developed independently./3/ In the field of religious studies the question takes a somewhat different form: is there a "perennial philosophy," a "primordial tradition," however obtained, shared by major religious thinkers of antiquity?/4/ Or are there deep divisions and genuine diversity in religious belief systems? To all of these and related questions early Chinese cosmological views are directly relevant, and thus here, too, they should be of interest to us.

These answers are important—much more important than my highly adumbrated statements of them suggest—but they clearly evade the sceptical thrust of the question, which is directed toward the noun in the title of this anthology: *cosmology* is now the province of astronomers and astrophysicists, and for any cosmological account to have a serious claim to our attention it must employ, in addition to "pulsar" and related terms, other unusual expressions now used technically, like "big bang," and "black hole"; terms and expressions for which there are not even rough equivalents in the lexicon of archaic Chinese. In *this* sense, then, while the study of early Chinese cosmology may teach us much, it cannot teach us anything about the structure of the universe; it cannot, that is, teach us anything about *cosmology* in the modern sense.

I want to maintain that it can, because our *concept* of cosmology must needs affect our cosmology, and read in the right way, the essays herein can enable us to bring that concept to the surface from the mental deep for examination. To be sure, there are no terms in archaic Chinese analogous to most of the basic terms in the parlance of contemporary astronomy and astrophysics, and thus we are almost surely going to be disappointed if we approach these essays asking, "To what extent do these Chinese writings suggest answers to cosmological questions which perplex us?" Neither the Shang oracle bones, nor the *Huai Nan Tzu*, nor anything written in between and described herein will deepen our scientific understanding of, say, what we refer to as the Crab Nebula.

On the contrary, readers will encounter in the pages to follow such terms as "alternation," "bipolarity," "hierarchy," "order," "complementarity," and similar expressions. At first reading these terms will probably not fit in with the descriptions we would give of our cosmological views. But we must read more carefully, because the contributors demonstrate in their several expositions how well these and related terms fit with each other, and how they can be seen to be fitting with respect to the world "out there." Again: the authors, individually and collectively, offer here a coherent, convergent and comprehensive (early Chinese) cosmology, and they invite us to enter into it, necessitating another approach to these essays, namely by asking "To what extent do these Chinese writings suggest that we might ask very different cosmological questions?"

Deep waters indeed, which if not probed carefully will become muddied. In the first place, I am *not* maintaining that early Chinese cosmology is *better* than ours, more nearly *correct* than ours, or, in some sense, *true*; any or all of the above may be the case, but are not my present concern./5/ Rather do I want to maintain (1) that the basic conceptual framework of ancient Chinese thinkers is different from our own; (2) that we can profit much by attempting to comprehend their conceptual framework because (3) we may learn much about our own conceptual framework, as well as theirs, by doing so. In other words, the cosmological texts examined below are not, in my view, merely a window through which we may see the conceptual framework of the ancient Chinese, they are also a mirror by means of which our own conceptual framework may be held up for reflection.

But can this be done? How can we examine our *basic* conceptual framework? It would, for example, be a difficult intellectual task to compare and contrast the basic conceptual framework reflected in the *Upanishads* with that reflected in the *Tao Te Ching*, Sanskrit and Chinese. Obviously, however, it would not be an impossible task, for we have a relatively *more basic* conceptual framework within which our assessments would be made, namely, our own; in English./6/ When one of the frameworks to be compared, on the other hand, is our own, in

English—and hence, we do not have access to anything more basic—it is not at all obvious that what I am suggesting is possible even in principle.

In a similar way this problem has increasingly exercised Western philosophers ever since Kant. The following quotation from Imre Lakatos, although written in a specific context, can nevertheless serve as a succinct summary of much 20th-century Anglo–American philosophy:

> The early Wittgenstein of the *Tractatus* rediscovered the fact that we look at the world through the spectacles of our conceptual frameworks, which are expressed in our language. This time-worn idea . . . is trivially true. Now there are apriorists who aim at manufacturing perfect spectacles. Others, like Popper, aim at appraising the relative merits of different spectacles. But the later Wittgenstein denied that one can distinguish the quality of one pair of spectacles from another: all that one can do is to clean the spectacles one has got./7/

The spectacles metaphor—originally Popper's—may at first appear to be appropriate for clarifying my argument: is my claim that we should (can) put on the Chinese spectacles to see what we might see? The major problem with the spectacles metaphor is that they stand for beliefs, and we can certainly place our beliefs in front of us for examination and question. In just the same way, with effort we can see our spectacles—especially if they need cleaning. With even less effort we can see the frames of our spectacles, the periphery of our conceptual framework, which may well change with philosophic or scientific fashion just as we change lens frames periodically with other changes of fashion.

I want to go further, however, and suggest that a sympathetic reading of the Chinese texts and commentaries will make more accessible to us the *presuppositions* of our beliefs, a more difficult but potentially more liberating (exciting, progressive, humane) activity. My own metaphor, then, is contact lenses. They are much more diifficult to put on and take off (out) than glasses, we cannot see them no matter how hard we try, and we certainly cannot see their boundaries. We only become aware of them when some foreign matter enters, and/or our vision gets blurred.

Let us now return to the concept of cosmology. This anthology is properly titled, because cosmology is the study of the way the components of the universe are arranged and ordered. Together with cosmogony (the study of the origin of the universe) and ontology (the study of what there is in the universe), a special field of philosophy is demarcated, which, thanks to a happenstance in the editing of Aristotle's writings we have come to call "metaphysics." Along with the other fields of philosophy—epistemology, logic, ethics, aesthetics, etc.—this categorization, apart from the specific sciences, encompasses all that human beings might wonder about. And at this level of generality we are guaranteed objectivity, neutrality, and clarity of vision. Specific philosophies may of course lose any or all of these qualities, but the categorization itself keeps everything open.

Or so we *presuppose*, until we consciously examine the categories. I do not think it would be correct to say initially that we *believe* this categorization, for it is within it—an integral part of our basic conceptual framework—that our cosmological, ethical, and other philosophical speculations take place, and the results thereof find expression. But this categorization must needs affect the form and thrust of the questions asked, and consequently, of the answers given.

Suppose we inherited another categorization, a "fourfold wisdom path": (1) knowledge of the patterns of the heavens; (2) knowledge of the things of this earth; (3) knowledge of human beings; (4) knowledge (true wisdom) of how to lead one's life.

I do not want to suggest that this proffered categorization is Chinese (it isn't), nor that it is better than the one we have inherited. It is, however, a possible .categorization, and sufficiently dissimilar from our own to suggest that if it were *presupposed*, the philosophical questions asked might well be different, and to suggest also that the questions we ask, some of them at least, might not even arise. It is by no means easy to appreciate this point. We might, for example, want to ask where aesthetic concerns would fit into the fourfold wisdom schema: under knowledge of human beings? Or under knowledge of the things of this earth? Might we need a hybrid category? But again, aesthetics is *our* categorization; aren't these questions *question-begging*?

In a way, then, the title of this book *may* be misleading. The ancient Chinese authors of the texts discussed in the following essays wrote what *we* would categorize as (naive, probably) cosmology, but the essays will reveal to the careful reader a different categorization. Irritants may make us painfully aware of our contact lenses. Cosmology may blur into epistemology. We are inclined, for instance, to distinguish knowing *how*—to swim, ride a bicycle, etc.—from knowing *that* (some fact); and the latter may be further distinguished as between "knowledge for its own sake"—more venerated than practiced—and "instrumental" knowledge, that is, knowledge of cause and effect, knowledge of how to control and manipulate nature broadly defined.

The seven scholars represented below, however, all make explicit that the ancient Chinese studied the cosmos significantly in order to learn how to pattern their individual and societal lives. Is this knowing how or knowing that? Is it instrumental or non-instrumental knowledge? Or is the issue not epistemological, but within the category of political theory? Are there more insights to be gained, perhaps, by not asking such questions?

At the very least we should endeavor to bracket them, and enter into the Chinese universe in the pages to follow. As will become clear, that universe spoke to the Chinese of long ago. It told them what was good and what was not, what made for harmony and what produced cacophony. *Our* universe is silent. The billiard balls, from quanta to quasars, move about in ways that can be made intelligible, but they do not speak to us; their motions and patterns cannot have personal or social significance for us, for within our conceptual framework they have no significance whatsoever.

Thus, if we do not make an effort to bring our basic presuppositions to the fore, the ancient Chinese writings encountered herein will seem naive, primitive, or magical at best, and at worst as examples of that queer conceptual entity some are wont to call "Chinese logic." I doubt that either of these interpretations—really little more than name-calling—have much to recommend them. At the same time I also doubt that the ancient Chinese views are the be-all and end-all account of the universe and our place in it. What I do not doubt is that the ancient Chinese texts, so ably brought to life in the following essays, can play a significant role in opening new vistas on the universe, and on those cosmically small but important creatures who have for so long wondered about it, and their place therein./8/

Perhaps we can even learn to see—without spectacles, without contact lenses—things we have not seen before.

III

It remains to give proper acknowledgements, offer appropriate thanks, name names; all of which it is a privilege to do.

The American Council of Learned Societies provided the major source of funds for the Harvard Workshop at which these papers were first presented, and they provided needed intellectual and moral support as well; all Workshop participants are deeply grateful to the ACLS. Harvard University, the Society for Asian & Comparative Philosophy, and the International Exchange of Scholars (IREX) also made important contributions of money and support for the Workshop. (For additional details on the Workshop and the participants, see the Introduction to the earlier companion volume to this one.)/9/

Workshop papers, with a précis, were submitted to a number of publishers consecutively, all of whom agreed that the papers were significant for scholarship, *but* that their publication would probably not be a commercial success. Against this background readers will appreciate a generous expression of gratitude to the editorial board of the Scholars Press in general, and to Ray L. Hart and Robert P. Scharlemann—past and present editors of the *Journal of the American Academy of*

Religion—in particular, for insisting that worthwhile materials be made available despite their problematic profitability. Robert Scharlemann has also been of great assistance in preparing these papers for the press.

I am indebted to Chen Jun for the elegant calligraphy which graces these pages; and to Constance Rosemont, for undertaking the difficult task of compiling a competent index.

For the delay in this book's being published, and for all technical errors that may be yet found in it, the editor accepts responsibility, and offers apologies.

Henry Rosemont, Jr.

NOTES

/1/   Compare the radically different accounts given of the world-view of 19th-century Vietnamese peasants by James C. Scott, *The Moral Economy of the Peasant*, Yale Univ. Press, 1976; and Samuel Popkin, *The Rational Peasant*, Univ. of California Press, 1979.

/2/   See, for example, Hans-Georg Gadamer's "Idea and Reality in Plato's *Timaeus*" and Karl Popper's analysis of the same work. The former article is in *Dialogue & Dialectic*, trans. P. Christopher Smith, Yale Univ. Press, 1980. Popper's critique is in the first volume of *The Open Society & Its Enemies*, 5th revised edition, Princeton Univ. Press, 1966, esp. 247–53.

/3/   Some push back *very* far: Giorgio de Santillana and Hertha Duchend, *Hamlet's Mill*, Gambit, Inc., 1969.

/4/   A recent account is in Huston Smith's *Beyond the Post-Modern Mind*, Crossroad Pub. Co., 1982.

/5/   Writings in this vein include F. Capra's *The Tao of Physics*, Shambala, 1975, and his more recent *The Turning Point*, Simon & Schuster, 1983. See also Gary Zukav's *The Dancing Wu Li Masters*, Morrow, 1979. Capra's views are examined in an essay by Stephen Jay Gould in *The New York Review of Books*, March 3, 1983.

/6/   Some philosophers, of course—not alone the later Wittgenstein mentioned above—would dispute this claim: W. V. Quine, *Word and Object*, MIT Press, 1960; Thomas Kuhn, "Reflections on My Critics," in I. Lakatos and A. Musgrave, eds., *Criticism and the Growth of Knowledge*, Cambridge Univ. Press, 1970; Paul K. Feyerabend, *Against Method*, Verso, 1982. And others. Much of the scepticism, however, except in the case of Feyerabend, stems from a view of natural language—and hence translation—that I believe mistaken, which I will attempt to show in *On Comparative Philosophy*, in preparation.

/7/   Imre Lakatos, *Mathematics, Science and Epistemology*, ed. J. Worrall and G. Gurrie. Cambridge Univ. Press, 1980, 229.

/8/   I have not touched explicitly on what is probably the deepest epistemological issue of all, Cartesian dualism. Routed as a philosophical system, the subject–object split has nevertheless become so embedded in our language that it may appear futile to think we may come to accurately comprehend views originally expressed in a language in which the dualism may not be evidenced at all. This scepticism may turn out to be warranted, but (a) it underestimates, in my opinion, the power of language—see footnote 6—and (b) such scepticism can all too easily be used as an excuse by philosophers for ignoring the intellectual heritages of non-Western civilizations.

/9/ *Studies in Classical Chinese Thought*, ed. Henry Rosemont, Jr., and Benjamin I. Schwartz, Scholars Press, Thematic Studies Series of the *JAAR*, 1979.

Late Shang Divination:
The Magico-Religious Legacy

David N. Keightley

Introduction

The figure of the high god Ti, ancestral rituals and sacrifices, the central value of filial piety, burial practices, the luni-solar calendar, the language and writing system—most aspects of Chou culture have their Late Shang antecedents. And by acknowledging in such catalogs the debts the Chou owed the Shang we may push the antiquity of Chinese culture back by at least two or three centuries.

But culture is not just content, it is also form. It is not just thoughts, it is ways of thinking. As Geertz has noted, "Culture patterns—religious, philosophical, aesthetic, scientific, ideological—are 'programs'; they provide a template or blueprint for the organization of social and psychological processes."/1/ These culture patterns will be my concern in this essay. By studying the oracle-bone inscriptions of the Late Shang elite— the inscriptions, that is, from the reigns of the last eight or nine Shang kings (ca. 1200–1050 B.C. in my view)/2/—I hope to suggest ways in which the mental habits, psychological dispositions, and logical expectations which the Chou inherited from the Shang as part of an authoritative tradition influenced the nature of Chou thought and culture itself. By understanding the magico-religious origins (or, more exactly, the earliest documented instances) of some of the mental constraints within which the Chou operated, and which gave their thought much of its great power, we may gain fresh insights into the nature of Chou culture. The oracle-bones of Shang provide a fresh perspective from which to view the ancient Chinese, a fresh perspective from which to view the metaphysical-moral gulf, frequently unappreciated, that separates them from ourselves./3/

It is not easy to make a distinction between the methods and goals of magic and religion, and there is no reason to think the Shang did so. For analytical purposes, however, I will assume that religion involves constraint, manipulation, and control of supernatural forces. Religious practices are far more expressive, making explanatory statements about the true nature of the world; magical practices are more instrumental./4/ But both kinds of activity are symbolic, a term which refers both to their nonrational (*not* irrational) character and to their suprarational power. As Huizinga has put it: "From the causal point of view, symbolism appears as a sort of short-circuit of thought. Instead of looking for the relation between two things by following the hidden detours of their causal connections, thought makes a leap and discovers their relation, not in a connection of cause or effects, but in a connection of signification or finality."/5/ And, as Gertz has pointed out, the peculiar power of symbols "comes from their presumed ability to identify fact with value at the most fundamental level, to give to what is otherwise merely actual, a comprehensive normative import."/6/

The oracle-bone inscriptions are not optimum documents for assessing the full range of Late Shang mentality, yet they have much to recommend them. Where divination flourishes, its logic and assumptions are not likely to be at variance with those of the rest of life./7/ Further, the Late Shang divined most aspects of life; the belief that the future could be divined was undoubtedly as real to the Shang as any other symbolic aspect of their culture. And it is clear, too, that the Shang elite devoted great amounts of time and energy to pyromancy./8/ For all these reasons, therefore—the "normal," comprehensive nature of their logic and assumptions, the scope of their concerns, and the attention paid to them—we are justified in studying the oracle-bone inscriptions for significant signals about the Late Shang worldview. In what follows, I shall first attempt to characterize significant aspects of Shang divination and its underlying conceptions; I will then consider their possible Chou legacies.

Late Shang Divination

The Rationality and Clarity of the Divination Charges

Shang divination was presumably a symbolic activity. It was entwined with magico-religious rituals that were both expressive and instrumental, and which shaped Shang views of reality and religious forces at the same time that they embodied the attempt to control them./9/ Rituals presumably rendered the divination efficacious (their magical, instrumental function); at the same time, they encouraged the view that all life could be and should be divinable (their religious, expressive function)./10/

There is nothing manifestly symbolic about the inscribed divination charges, however, which were recorded in straightforward, administrative prose—"it will rain"; "it will not rain"; "the king should ally with this tribe"; "the king should attack that one"; "the king's dream was caused by such-and-such an ancestor"; "there will be good harvest," etc. The function of the charges might vary (see pp. 14–16 below), but they were all recorded "in clear." An-yang was not Delphi, where the words of the Pythia had to be translated, frequently ambiguously, by priests./11/ Shang divination was "cool" and ordinary, uninspired and rational.

The fact that, in the reign of Wu Ting (Tung Tso-pin's period I), /12/ the charges were frequently recorded as complementary charge pairs in the positive and negative mode ("Fu Hao's childbearing will be good/Fu Hao's childbearing will not perhaps be good")/13/ reinforces our sense that the Shang diviners made contact with the supernatural not by relying on trance or the suspension of normal patterns of thought, but by proposing mundane, yes-no, positive-negative, alternatives. The supernatural powers were given little opportunity to inspire new solutions. Man proposed his simple, pedestrian (i.e., human) alternatives; the supernatural could only choose between them./14/ The "coolness" of the divination forms was congruent with the "coolness" of the language of the charges. It is true that the use of positive-negative modes may also have had symbolic significance, reflecting a metaphysical sense of yin-yang style complementarity, in which alternatives were inextricably entwined with one another,/15/ but the fact remains that few if any charges were couched in this bifurcated mode after the reign of Wu Ting, and that all charges were precise and limited in the options they offered for consideration. The powers could only dispose in ways that man had already proposed.

The Order of the Divination Charges

There is an *ordnungswill* strikingly evident in the motifs and styles of Shang art, whose geometric patterns and highly formalized animal shapes are eloquent witness to the dominance of abstract order in the Shang worldview; such art was the pictorial and plastic expression of the structures of Shang thought and action, an expression of one of the *daimons* that gave the Shang their power and confidence.

The form of Shang divination, as well as its content, was also marked by a profound sense of order that must be taken as expressive of some fundamental view of reality and man's relation with the supernatural. Hollows were chiseled or bored into the backs of scapulas and plastrons with great care and effort so that the *pu*-cracks, which appeared on the front when the diviner burned the hollows, formed in their turn a series of preordained patterns. Unlike the free-form pyromancy of other cultures, where the bone might be thrown into the fire, or the heat applied to any point of the unprepared bone surface, there was nothing random about the pyromantic cracks of the Late Shang. Just as the clear, positive-negative charges imposed order on the powers, the hollows imposed order on the cracks. No crack could appear where the Shang diviner did not want it to. The powers could not reveal themselves in unexpected ways. The supernatural responses were rigorously channeled.

A similar sense of order may also be discerned in the content of divinations about ritual and sacrifice. Ancestors whose jurisdictions might overlap were ranked in a junior-senior hierarchy and sacrifices were offered to them according to an elaborate, fixed schedule. Divinations were concerned with offering the right number of the right kind of victims to the right ancestor on the right day. Even in the reign of Wu Ting, when the system of sacrifice was still developing, there was little opportunity for invention or spontaneity. By the reigns of Ti Yi and Ti Hsin (Tung Tso-pin's last period, period V), the Shang kings no longer sought approval for scheduled sacrifices. They simply informed the powers that the sacrifice was being offered and expressed the wish that there would be no fault or misfortune. Uncertainty had been replaced by pattern and order. System now gave as much reassurance as the auspicious approval of the powers had formerly done. And the powers themselves, presumably, were thought content with this solution. The existence and efficacy of order, particularly where time was concerned (see pp. 17–18 below), was a paramount article of Late Shang faith.

The Prophetic and Transcendental Worldview

All divination, to the extent that it predicts the future, or seeks to reveal the true, supernatural meaning of past events, is transcendental. It is the special quality of Shang prophecy—so different, say, from the ecstatic, personal, and frequently critical and unsettling prophecy of the Old Testament—that concerns me here. The Hebrew prophet announced a message or vision he had received from God; the Shang diviners announced their message to the spirits./16/ The difference is striking and critical.

During the reign of Wu Ting, Shang divination served several related functions. The charges and cracks were used to identify spiritual forces, to document the approval of the spirits, to pray for blessings and assistance./17/ The accompanying rituals energized the divinatory act. Faith in the efficacy of the procedure was also maintained by what I have called "display inscriptions" which recorded in large, bold calligraphy the accuracy of the king's forecasts. In a typical example, the king divined about whether there would be disaster or misfortune of some sort in the coming ten-day week; examining the cracks, he prognosticated that there would be a disaster; and, *mirabile dictu*, the verification records that disaster, such as an enemy incursion or a lunar eclipse, did occur as predicted./18/ Ostensibly at least, unknown events were assigned ominous significance *before* they had taken place./19/ Future events, explained before they occurred, demonstrated the efficacy of the king's mantic powers.

In addition to what we may call these "mantic, pre-facto portents," the Shang diviners under Wu Ting were also concerned with more traditional, "inductive, post-factum portents"—as in divinations to discover which ancestor was responsible for a sickness or dream. In these cases, the past was explained so that some remedial, sacrificial action might be taken.

Divination as Magic

The bulk of Shang pyromancy, however, was not directed toward this kind of prophecy. "Divination" did not simply have the sense of forecasting or retrospective discovery that it has for us. After the reign of Wu Ting at least, Shang divinatory concerns were not generally focused on forecasts of unknown misfortunes that might befall the king (i.e., pre-facto portents), or on explications of specific misfortunes that had already befallen him (post-factum portents). Most of the divination charges stated to the spirits what it was that the king proposed to do— offer a sacrifice to an ancestor, attack an enemy statelet, hunt at a certain place on a certain day—and attempted to discover whether there would be "assistance," or "no misfortune," or "no regret," or "no disaster," if he indeed did so.

It is likely, however, that the Shang, in divining these statements, sought more than the pyromantic imprimatur of the powers. The evidence is not as strong as one would wish, but it seems reasonable to suppose that much Shang divination was incantatory in nature. When good and bad alternatives were proposed, the charge recording the bad or undesirable alternative usually contained the particle *ch'i* 其 which, it seems, weakened the force of the charge (and whose presence I indicate by a "perhaps"); such undesirable charges might also be weakened by being abbreviated./20/ Thus, a period I divination of the form "We will receive millet harvest" (right side) and "We will not perhaps receive millet harvest" (left side) should be regarded not simply as an attempt to *discover* whether or not there would be a good crop, but as an attempt to *ensure* that there would be./21/ The possibility that there would not be was weakened by the *ch'i* of the negative charge. The very fact that the charges were incised with great labor into the bone—that incising not being related, apparently, only to the desire to preserve the record for posterity (see p. 21 below)—may well have been due to the desire to "fix" the divinatory spell, "carve" it into the future, and render it efficacious./22/

By period V, undesirable alternatives had virtually been excluded from the divination charges, which had become resolutely optimistic and desirable: "If the king hunts at Sang, going and coming there will be no disaster"; "In the (next) ten days there will be no disaster"; "This evening it will not rain"; "The king entertains and performs the *sui*-ritual; there will be no fault."/23/ These charges may be regarded not just as religious applications for spiritual approval, but as magical charms to ensure that there would indeed be no disaster, no rain, no fault. The charges

were not interrogative; they were assertive and impelling, informing the
powers what the Shang were doing, what the Shang wanted, and
attempting to ensure that the results would turn out as desired and forecast./24/ If, as has been said, "magic demands, religion implores,"/25/
the late Shang charges may be seen as primarily magical and instrumental in intent, though, as we have seen, the rituals associated with them
undoubtedly had their religious, expressive function. If one function of
Shang divination was to encourage the occurrence of what had been
forecast, then when what was forecast did happen, the event served in
effect as a pre-facto portent, validating once again—though less explicitly than the display inscriptions of Wu Ting's reign had done—the
king's role as seer and archimage. It should also be stressed that the king,
as he appears in the inscription record, was generally an infallible prophet (and thus magus). Few if any verifications record that his forecasts
or charms were proven wrong./26/

The logic of the divination charges may also be characterized, like
the view of reality which they divined, as magico-religious. That is to
say, the correlation between two (or more) events linked together in a
charge was spiritual, not human; the cause-and-effect depended on spiritual approval, not on empirically verifiable actions. It depended, in Huizinga's terms, on "a connection of significance or finality." Thus, to
return to an example already cited, the charge, "If the king hunts at
Sang, going and coming there will be no disaster," presents two ideas,
juxtaposed by the act of divination itself. By calling the attention of the
spirits to the hunt, and by performing the divinatory ritual, the king
tried to ensure that there would be no disaster. But the result was contin-
gent upon the approval of the spirits. In the most general terms, therefore, Shang divination, in which the charges proposed what were, in
effect, pre-facto portents (p. 5 above), was of the form "We do A and B
will happen." The post-factum charges merely reversed the sequence: "B
has happened and A did it." The "and" which correlated the events
existed at the pleasure, and expressed the will, of the spirits. That pleasure, that will, could be divined by man and could be influenced or
impelled by the divination charges and prayers, and by the attendant
sacrifices. Man could propose his actions and the result that he wished;
but it was the spirits who could make, or not make, the connection
between the two.

By period V that connection was generally taken for granted; the perfunctory nature of the routine incantation suggests there was little doubt,
or little concern, that the spirits would oblige. Though still primarily magical in its intent, divination was becoming more religious./27/ But for all its
routineness, the cause-and-effect by which the world worked was conceived in symbolic terms.

The Categorical Unambiguousness of the Divination Charges

As I have already suggested (pp. 13–14 above), the divination charges were couched in terms that left little freedom for spiritual inspiration or invention. Under Wu·Ting the complementary charge pairs generally offered only two options: "It will rain/it will not perhaps rain," "We will receive assistance/we will not perhaps receive assistance," etc. By the late periods, a single option was the norm: "It will rain," "We will receive assistance." And to such restricted divinations, the cracks could give only three responses—auspicious, inauspicious, or neutral./28/ Some gradations were provided for—crack-notations might, for example, read "highly auspicious," "greatly auspicious," or "less auspicious"—but generally the world was conceived in sharply delineated alternatives, which either prevailed or did not. There was no room for subtle interpretations, paradoxical responses, or deceptive meanings concealed in obscure prognostications./29/ In this regard, Chaucer's lines,

> For goddes speken in amphibologies,
> And for o soth, they tellen twenty Lyes,

could hardly be applied to Shang divination. The ambiguities so characteristic of many other systems of divination, such as the Delphic one,/30/ did not plague the mind of the Shang diviner, who saw— or at least recorded—a world in black-and-white, positive-negative, auspicious-inauspicious terms. Such a cast of mind is congruent with the strong sense of patterning and classification to which I have already referred (pp. 13–44 above).

The Flexibility of Timeliness

Such a system of divination—with charges and responses rigidly conceived and narrowly construed—would seem, to us at least, to have been a clumsy and unsatisfactory way for dealing with the supernatural. The interstices, the "greys" of human experience, unaccounted for by such a system, may well have been handled by other kinds of divination of which we have no Shang record; the Yi-ching, for example, with its enigmatic aphorisms, may well have arisen to fill this need (though even the Yi-ching, it should be remarked, presents a limited number of situations and responses).

The Shang diviners of Wu Ting's time, however, obtained some flexibility by the strong emphasis they placed upon timeliness. Kuei-day divinations about the fortune of the coming ten-day week; divinations about performing sacrifice on the next chia, yi, or ping day and so on; divinations about rain from this day to that, or in this month; prognostications about childbearing or disaster that were verified on a later day—the endemic interest in time questions of this sort reveals what must have been a vivid and intense concern with time future, time forecast, time sanctioned by the approbation of the spirits. Time, as it appears in the inscriptions, was religious. The *kan-chih* calendar recorded in regular order the days on which rituals and sacrifices were offered to the ancestors (who were themselves named by the *kan* stems). The fundamental importance of religious, or cultic, time can be seen from the way in which late Shang inscriptions, on both bone and bronze, routinely date events in terms of the ritual cycle. But it is not so much the theology of time that concerns us here; it is the temporal dimension of Shang divination itself.

The fact that nearly every divination preface records the *kan-chih* date indicates that the time when the bone was cracked had significance. Presumably, the crack indicated the disposition of spiritual forces at that time; the prognostications, attached to the dated charge, applied only at that time. This interpretation is supported by the fact that the Shang would, on occasion, divine the same topic over a series of days; the repetition suggests that new results may have been expected as the day of divination changed./31/ Further, those cases in which the prognostication is more detailed than the original charge carved on the bone or shell—e.g., charge: "Fu Hao's childbearing will be good"; prognostication: "If it be a *ting*-day childbearing, it will be good. If it be a *keng*-day childbearing, it will be extremely auspicious"/32/—indicate that the Shang diviner assigned a specific day tò each crack as he cracked it, even though such putative subcharges ("A *ting*-day childbearing will be good" . . . crack! . . ."A *keng*-day childbearing will be good". . . crack! . . . , etc.) were not recorded on the oracle-bone (see n. 28).

This concern with timeliness was characteristic of the divinations of Wu Ting, who reigned when the sacrificial schedule was still being formulated, when the date of each sacrifice might still be submitted for spiritual approval. By period V temporal flexibility had been lost. The ritual schedule was now rigidly formulated; the day on which a particular ancestor would receive sacrifice was already established; the divination was no longer concerned with determining the auspicious time, but only with announcing that the sacrifice would take place as scheduled.

This loss of temporal flexibility was also related to a reduction in the temporal scope of the charges. Under Wu Ting, the diviners might on occasion propose sacrifices twenty days in advance; but by the late period, the range of the divination never exceeds the ten-day week, and no sacrifices were divined in advance./33/ Timeliness was still essential, but it was no longer problematical. Time had been ordered, but a significant option for assessing degrees of, and changes in, spiritual approval had been lost—not lost, perhaps, to Shang culture (for other systems of divination may have emphasized the option of timeliness), but lost to the pyromantic record.

The Late Shang-Chou Transition: A Hypothesis
    The scope of Shang divination had dwindled remarkably by period V.
Divinations about the sacrificial schedule, the ten-day period and the
night, and the hunt were still performed in great number, but dreams,
sickness, enemy attacks, requests for harvest, the issuing of orders, etc.,
were divined far less frequently than they had been in period I, if at
all./34/ This reduction, associated with the disappearance of positive and
negative charge pairs, with the disappearance of inauspicious prognostications, and with the reduced temporal range of the forecasts, suggests that
a major shift had taken place in magico-religious theology. We must
assume, not that the Shang were no longer troubled by such questions as
successful childbirth, victorious alliances, or the significance of dreams, but
that other systems, like that represented by the *Yi-ching*, grew up to handle
them. In period I, the king's toothache, for example, was regarded as the
result of a spiritual curse; it was treated by determining, through divination, the ancestor responsible for the sick tooth, and by discovering which
sacrifice, on which day, would lead to a cure./35/ The kings of period V
presumably still suffered from toothache. The fact that they no longer
divined about it suggests that a new explanation for curing toothache had
been developed. Where a toothache (or drought, or accident, etc.) had formerly been considered a post-factum portent, symbolic of ancestral dis-
pleasure and requiring pyromantic spell and placatory ritual, it may now
have been symbolic of something else.
    I would suggest that, by the Late Shang, misfortunes were increas-
ingly interpreted, not as the result of arbitrary or malicious actions taken
by dead ancestors, but as the result of actions taken by their living
descendants, explainable by a consistent set of religious values. A sickness
would still have been a post-factum portent, but it might have been
viewed as the result of human failing, a deficiency, at first, perhaps, in
the attention given to the ancestors, and thus, by extrapolation, a deficiency in "virtue." What the content of that "virtue" may have been cannot be deduced with certainty from the oracle-bone inscriptions, though
I would suppose, given the great attention the Shang paid to ancestral
sacrifices, that virtue was related to ancestral sacrifices and, by extension,
to filiality. Ethical action—the way one man treats another—would not
yet have assumed major importance in the belief system, but the initiative in the relationship between man and the spirits, the responsibility
for making the world work, would have been passing to man.
    If these speculations are correct, the dwindling scope of the oracular
record, together with the increasingly perfunctory nature of its routine
incantations, may be seen as part of a shift from magic towards religion,
from charm to prayer. In this view, the Late Shang came to rely increasingly on religion (human intercession) rather than magic (divinatory
coercion) to deal with the shocks of life. To the extent that such religious

action, which was undoubtedly highly ritualized, replaced pyromancy, the scope of Shang divination was reduced accordingly. But to the extent that such action was the functional analogue of divination, it continued to have a mantic, magico-religious quality. Virtuous conduct, whatever its exact nature, would have been efficacious and desirable, not in its own right, but because of its symbolic, transcendental character. It was still a symbolic response not to misfortune or sickness per se, but to what was still seen primarily as a portent. Virtuous action had replaced actual pyromancy. But virtuous action itself was still an attempt to impel events by the old magical logic we have discerned in pyromantic divination./36/ And in both cases, divinatory or ethical, rituals were essential to proper action.

The Legacies, Metaphysical and Ethical

I have argued elsewhere the ways in which Shang ancestor worship may help explain the high value placed on bureaucratic modes of thought and action, the way in which the pervading sense of religious obligation was congruent with, and encouraged the development of, a work ethic based on obligations and duties rather than rights./37/ In what follows, I shall suggest some of the ways in which the magico-religious styles of Late Shang divination continued to affect the mental habits of later Chinese thinkers, especially in the area of metaphysics and ethics.

The fact that Shang divination appears to have involved no shaman-istic flight to other realms, and that it employed normal language, normal consciousness, and normal systems of choice, may be related to the immanent metaphysics of later Chinese thought./38/ Just as heaven and man were later viewed as part of one existential continuum, so were man and the ancestors in Shang times; communication between them required no disruption of ordinary modes of existence. There is little evidence that the powers (and I am speaking here primarily of the ancestors, to whom the bulk of Shang religious attention was turned) were conceived as "wholly other."

The Shang did not reject transcendence—their world was quintes-sentially transcendent; it derived its significance from an endless succes-sion of portentous events whose meaning could be discerned, and whose outcome could be influenced, by divination. Shang divination implied a unitary but transcendental metaphysics in which reality was pregnant with, inseparable from, immanent "signification or finality." Reality had to be sought within phenomena, an approach strikingly different from that of the Western philosophers who have sought reality outside of phenomena./39/ But phenomena, if they were real, thus belonged to a larger system of sense and order. The later Confucian belief that a simple historical record (like *Ch'un-ch'iu*), or a man's ordinary actions, should be judged in transcendental, moral terms, the conviction that reality could be explained by *yin-yang* or *wu-hsing* theories, derives from this same Shang tradition which did not view the world in value-neutral terms, but as a series of phenomena whose true significance had to be divined. History and human conduct replaced divination as a means of foretelling the future;/40/ past events and actions could be read as the Shang diviners had read the cracks. Didactic history was a record of pre-facto portents.

The whole tradition—which we may crudely characterize as the *chin-wen* view of reality—that has attempted to discern hidden meanings in mundane texts and events derives from, and takes strength from, the divinatory theology of the Shang elite. A view of reality, so long held and sanctioned by religious belief, would not easily have been discarded. Even in recent times, the desire to explain seemingly insignificant acts in terms of consistent, moral-political theories, to divine what a man's acts really mean, what his motives are, and, in so doing, to bring about changes in his behavior, is a characteristic feature of Chinese political culture. The explanatory power of the Marxist-Maoist portents ("Use the past to serve the present"), the political power of the *ta-tzu-pao*, is not unrelated to the magico-religious power of the portents and the display inscriptions carved on the oracle-bones of Shang.

For it is likely that the Shang divination records served a rather different function from those kept, for example, in the ancient Near East. There, "the purpose was clearly to record experiences for future reference and for the benefit of coming generations."/41/ Shang inscriptions, particularly the display inscriptions (see n. 18), may have served such a purpose for a while, but this was not, I suspect, the primary motive for incising them rather than merely writing them with a brush. The fact that many oracle-bones were stored underground with a variety of other Shang "junk" suggests they were not recorded for posterity./42/ And if the incised divination record itself was thought to have some magical power to help bring about what had been divined (see p. 15 above), then it may well be that this attitude affected later Chinese attitudes toward certain bodies of writing, which were not regarded simply as records or repositories of knowledge, but as accounts of what ought to happen, or what ought to have happened, i.e., accounts of not just facts, but "moral facts."

Similarly, the Shang concern with portents—with explaining reality in terms of the will of the powers—helps account for the fact that metaphysics did not develop as a major category of Chou thought. This was not because later thinkers were uninterested in metaphysical problems. But it is because, I would suggest, metaphysical questions were viewed as religious (and, by extension, moral) questions. The nature of reality was not an existential question but an essential question. Since reality was pregnant with meaning, the nature of reality was always conceived in terms of what it meant. Ethical speculation was a form of metaphysics. The study of reality without a study of its ethical significance would have had no meaning; it would have been like cracking an oracle-bone without caring about the result.

The categorical, yes-no, positive-negative, nature of the divination charges, another "cool" feature of Shang divination that allowed no surprise answers or inspired responses, also encouraged a view of reality in which there was only one right way; difficult or tragic alternatives were not the subject of philosophy. Fingarette has remarked on the absence of any well-developed, Western sense of choice in *Lun-yü*./43/ Some of the Confucian aphorisms, in fact, suggest the complementary forms of Shang divination pairs./44/ And one might remark too on the tendency to define ethical terms by their complementary opposites,/45/ as though the complementary restraints of period I were still at work, so that one could only define bravery by contrast with fear, *jen* 仁 by contrast with *yu* 憂 , etc.

I would suggest, too, that the Shang diviner's worldview, which dealt with the supernatural by compartmentalizing alternatives into discrete "bits"—day, ancestor, sacrifice, no assistance, etc.—encouraged a "coolness" of outlook which, like the "normalness" of the divinatory logic, may relate to the "failure" of abstract thinking to develop as markedly in classical China as it did, say, in classical Greece. The Shang diviner's concern was with specific forecasts, specific events, not with overriding insights or theories. Such modes of thinking were apparently so satisfying that the Chou philosophers, rational and secular though the content of their thought may have been, did not discard the "oracular" forms of earlier thinking, with their emphasis on particular case histories viewed in symbolic terms. No sense of universal law could develop from such a religious view, only one of hierarchy and case-by-case bargaining.

The impelling, assertive mode of the Shang divination charges may also bear on this point./46/ Even when positive-negative alternatives were offered to the spirits, the Shang diviners did not ask a question of the powers. Their contact with the supernatural was indicative, not interrogative. They did not ask, "Today, will it rain?" They stated, "Today, it will perhaps rain/Today it will not rain." Divination was not just a matter of determining what the spirits wanted; it was a way of telling the spirits what man wanted, and of seeking reassurance from the fact that the spirits had been informed. The moral certainty characteristic of Chou ethical thought, and related no doubt to the absence of any sense of original sin, is again a legacy of this confidence implicit in the Shang divination charges. Class interest may partly explain the high moral tone of some early Chou philosophy. But the impulse to tell others what to do, to let the *chün-tzu's* conscience guide everyone else's, surely stems in part from the belief that the enlightened man had the ability and charisma to issue oracular injunctions in this way.

Similarly, the faith in rectification of terms, the belief that good government involves the correct choice of words, and that reality will somehow come to accord with the words (*Lun-yü*, 13.7) may also be related to the magical, impelling character of the words of the divination charges. That the rectification was primarily ethical rather than metaphysical or logical,/47/ is fully consonant with the transcendental Shang worldview. In this way, too, the philosophers of Chou were intellectual descendants of the diviners of Shang./48/

The habit of not asking questions but of proposing answers, further evidence of a sense of self-confidence and optimism, served, as noted, to limit the kinds of answers that could be obtained. The answer, i.e., the charge that would be found auspicious, was known in advance. Symptomatic of the metaphysical incuriosity already referred to, this aspect of Shang mentality may explain the rather "close-lipped" nature of Confucianism, its disinterest in brainstorming, in imaginative speculation. One frequently has the sense in Chou philosophy that questions are only asked when the answer is known in advance. The *Kung-yang* or *Ku-liang* commentaries represent, perhaps, an extreme version of what we may call "nonexploratory interrogation" whose origins may be found carved on the oracle-bones of Shang. I would not want to say that, as a result of this legacy, later Chinese culture has been characterized by an indifference to asking questions. But I would suggest that questioning did not have the same value or quality that it had, say, in ancient Greece. It did not raise the expectation of new revelations. New questions—new in the sense that they trespassed into unexplored ground—were not asked, because new answers were not foreseen. If we accept the remark that "trespassing is one of the most successful techniques in science,"/49/ it may even be suggested that the "failure" of science to develop in China may have been related to the compartmentalized and limited nature of divination forms which gave no rhetorical or metaphysical support to the possibility of trespass. Shang divination charges, in the barriers they established, are metaphysical analogues of the ethical *li*.

Further, I would suggest that preference for chain-reasoning and analogical argument, again characteristic of much Chou thinking, may be related to the transcendental correlations of the oracle-bone charges. The content of Chou and Han thought was more ostensibly secular, but the way the world worked, the logical links between its elements, was still partly mystical and symbolic. There is, as Fingarette has noted, a significant, magical dimension to the thought of Confucius. "Is *jen* far away? As soon as I want it, it is here."/50/ Similarly, the famous series of sorites in *Ta-hsüeh*, ending "When things are investigated, knowledge

is extended; when knowledge is extended, the will becomes sincere . . .
when the family is regulated, the state will be in order; and when the
state is in order, there will be peace throughout the world,"/51/ belongs
to the same logical tradition as the "We do A and B will happen" of the
oracle-bone inscriptions. That B will happen is an article of magico-
religious faith. The organic, synchronous worldview of the Chou and
Han, with its emphasis on pattern and relation, on significant and moral
juxtaposition, owed its inspiration to, and was congruent with, the divinatory logic of the Shang./52/ The Chou and Han knew the world
worked this way because the Shang had known it before them.

Reasoning has been defined as "thinking enlightened by logic." In
India and Greece, logic developed out of techniques for refuting arguments. To Plato, reason was the messenger of the gods./53/ To the
ancient Chinese, quite clearly, it was something less exalted, less supernatural, less subject to inspiration; it was a problem-solving device, not
an inspirational one. Reason was developed not in the refutation of arguments but in the stating of wishes.

The instrumental character of Shang divination may also have left
its mark on the action-orientation of Confucian ethics. "Meng Yi Tzu
asked what filial piety was. The master said, 'It is not being disobedient.'"/54/ We see Confucius's mind acting like that of a Shang
diviner, especially when we consider the Chinese text: *Meng Yi Tzu wen
hsiao* 孟 子 問 孝 , "Meng Yi Tzu asked about 'Hsiao,'" i.e., he just proposed the word, the concept. Confucius, and one can almost sense the
Shang mind at work, defined *hsiao* positively and negatively, "It is being
obedient" and "It is not being disobedient," and then selected one of the
two alternatives. Waley, in fact, translates the master's response, *wu wei*
無違 , as an imperative, "Never disobey!" And this I find extremely suggestive. For I have referred to the two alternative inscriptions of Shang
complementary divinations as "charges."/55/ The term seems accurate:
since one or the other charge was to be the answer, it was, in many
cases, an order or instruction as to what should be done./56/

This feature of Shang divination may help account for the fact that,
in Confucian rhetoric, answers to questions are not only circumscribed
by the form of the question itself, but are frequently given as moral
imperatives, or at least as answers which have the force of a forecast (as
in *Tso chuan*) that should be implemented. Is involves ought. Here then
may be a further legacy of Shang divinination forms which implied not
only an answer, but action in accord with that answer. To divine, in fact,
was to envisage and to commit oneself to eventual action, to involve
oneself in the world. The characteristic action-orientation of Confucian
ethics is congruent with the action-oriented form of the Shang diviner's
charges and may have derived in part from mental patterns reinforced
by the customs of divination. It may be objected that divination the

world over is necessarily action-oriented. That is true. But there is an attitudinal difference between asking "Should the king hunt?" and stating "The king will hunt." The Shang king, by the very way he stated his charges, disposed himself to action in the way that a mere questioner would not have done.

The love of order and hierarchy, so characteristic of Shang art, Shang divination forms, and Shang ancestral theology, continued to flourish in later times. With the exception of the Taoists, the Chou thinkers were horrified by disorder./57/ The major goal of Chinese religion of imperial times was to realize the way of heaven by preserving a universal order./58/ The order, where social theory was concerned, could certainly be justified in secular terms. But the passionate attachment to orderly, hierarchical solutions may be partly explained in terms of a fundamental faith in order as a good, quite apart from its efficacy./59/ What the Shang loved, the Chou did not reject.

That Shang ancestor worship was essentially a family affair not only had metaphysical consequences, lessening any division between the human and the spiritual; it also meant that the hierarchies of the family tended to be imposed upon the larger world, whether of man or the spirits. And this "human" dimension to Shang religion had further consequences, I would suggest, for the character of later Chinese humanism. In the West, as Mote has noted, humanism developed in response to (or against) religious authoritarianism./60/ In China, on the other hand, it appears to have evolved smoothly from ancestor worship; magical care of the ancestors (ex-humans) leads to quasi-religious care of the parents (living humans), and may lead to ethical concern for other humans. And the humanistic values never lost the primacy attached to religious ones. But, on the other hand, evolving smoothly from, rather than against, religious belief, Chinese humanism was not anti-authoritarian, not anti-hierarchical. And this is one reason, perhaps, why Chinese social and political theory, even to this day, has encouraged the development of an "authoritarian humanism," benevolent, concerned with the social well-being of all, but dependent on, and administered by, a central, magico-religious father-figure, both priest and official—emperor, *hsien*-magistrate, sage, chairman—who confers assistance as the Shang ancestors did, whose role was religious as well as secular, and whose wisdom was mantic as well as conventional.

This linking of traditional political values and expectations to the "programs" of Shang divination is not fanciful. The "programs" and their legacies were not restricted to diviners and philosophers, but, as we have seen (cf. n. 6), may be taken to represent the mental attitudes of the elite as a whole, dominating their modes of thought and imagination. The oracular impulse lies deep in Chinese culture. One of the marks of the Confucian *chün-tzu*, the political-philosophical paragon from whom even a ruler could learn, was his ability to foresee and prognosticate.

"The *chün-tzu* takes thought of misfortune and arms himself against it in advance," says *Yi-ching*./61/ "It is characteristic of the most entire sincerity to be able to foreknow," says *Chung-yung*./62/ And parts of *Tso-chuan* may be regarded as a Confucian divination text in which various prescient ministers, reading, if you will, the "cracks" in a man's character, are able to forecast, usually with astonishing success, the man's eventual and, in Confucian terms, thoroughly deserved fate./63/ Many of the stories in *Tso-chuan* are analogous to the display inscriptions of Shang, complete with charge, prognostication, and verification;/64/ they record pre-facto portents which are always validated, and thus validate the status and ethics of the prognosticating *chün-tzu*. What makes a man's speech persuasive is his ability to speak with moral authority about the future. We find a similar emphasis in the myths of dynasty founders. It was one of the special powers of a new king to recognize a man's worth despite his insignificant birth or position, and before his worth was manifest to others./65/ In all these cases, the message was Confucian, but the medium, the forecasting trope with its magical over-tones, was a legacy of Shang. And it may even be suggested that aspects of Marxist-Maoist historiography (and of the orthodox tradition of the dynastic histories from which much modern historiography has not divorced itself) which found nothing new in the historical record, but only demonstrations of preconceived theory, belonged to the same tradi-tion of metaphysical, ethical, and historical·prescience.

Conclusions

There is plenty of evidence that Shang traditions and Shang culture were still vital a millennium after the fall of the dynasty. The states of Lu and Sung were regarded as repositories of Shang culture./66/ Confucius advocated riding in the state carriage of Yin./67/ Both Wang Mang and the Kuang-wu emperor enfeoffed a Yin heir./68/ Sacrifices to T'ang, the dynasty founder, only ceased with the start of the Later Han./69/ There are countless references to Shang customs in *Li-chi*. If the particulars flour-ished in this way, so did the ethos and worldview. Every idea, every pat-tern of thought, has its genealogy, and many of the mental habits central to Chou and Han culture can be traced back, as I have attempted to show, to the ideas and thought patterns of the Shang. There were evolutions and refinements, but the "set" of Chinese thinking had generally been estab-lished a millennium or so before the golden age of Chinese philosophy.

Chou secular culture may be conceived as a rationalized descendant of the magico-religious culture of Shang. The religious achievements of the one age shaped the secular concerns and solutions of the next. We see a shift from a magical, religious culture to a moral one, but the mo-rality that evolved, and the ways of valuing human actions, were not without strong magico-religious overtones. If "religion is a man using a divining rod, and philosophy a man using a pick and shovel," the tools of the Chou philosophers, the vigor with which they were wielded, and the trenches they followed, owed much to the implements of their predecessors who had staked out the ground before them.

The fact that the magico-religious assumptions of Shang culture still played such a large role in the "inherited conglomerate" of Chou and Han suggests the degree to which these assumptions must have satisfied social and psychological needs. Whether or not the Shang diviners forecast or shaped the future with notable accuracy—and we might reflect on the record of modern economic forecasters before reaching too harsh a judgment—Shang divination clearly "worked" to satisfy the cultural demands of its believers, worked so well that its underlying assumptions continued to play a role for a millennium and more. The continuities are remarkably strong.

The rationality and clarity, order and hierarchy of Shang divination; its prophetic treatment of a transcendental, portentous reality; its optimistic confidence that human action could impel religious favor; its belief that the human predicament could be diagnosed and managed on a case-by-case basis; its great attention to the timeliness of human action; its human confidence that man could propose solutions to the spirits, rather than vice versa—all these features thrive, in secular guises, in the thought and culture of the Chou. For the end could never be, as Norman Brown reminds us, "the elimination of magical thinking." The goal could only be "conscious magic, . . . conscious mastery of those fires."/70/ As we consider the philosophers of the Eastern Chou, or even of twentieth-century China, who dreamed and wrote of an ordered society, it is salutary to reflect that order itself, for all its pragmatic benefits, needs to be fueled by the fires of belief. The fires of Shang burned long; they may be smoldering still.

NOTES

/1/   Clifford Geertz, "Ideology as a Cultural System," in *Ideology and Discontent*, ed. David E. Apter (New York, 1964), p. 62.

/2/   For the definition of this historical period and its absolute dates, see David N. Keightley, *Sources of Shang History: The Oracle-Bone Inscriptions of Bronze Age China* (Berkeley, 1978), pp. xiii, 171–76. Other chronologies have been proposed. Tung Tso-pin, for example, would include twelve kings in this period for which we have inscriptions, and would date it to 1398–1112 B.C. (ibid., p. 226).

/3/   Frederick W. Mote, "The Cosmological Gulf Between China and the

West," in *Transition and Permanence: Chinese History and Culture*, eds. David C. Buxbaum and Frederick W. Mote (Hong Kong, 1972), p. 6, has referred to "the Western failure to understand the basic nature of the Chinese world view." Different cosmologies, as he suggests, imply different views of reality, and they in turn are related to different ethical conceptions. It is the ethical-metaphysical conceptions that concern me in this essay.

/4/   Based on the discussions of: J. Goody, "Religion and Ritual: The Definitional Problem," *British Journal of Sociology* 12 (1961), 158–59; M. Fortes and G. Dieterlen, eds., *African Systems of Thought* (London, New York, Toronto, 1965), pp. 21, 24–25; John Middleton, ed., *Magic, Witchcraft and Curing* (Garden City, NY, 1967), p. ix.

/5/   J. Huizinga, *The Waning of the Middle Ages* (Garden City, NY, 1954), p. 203.

/6/   Clifford Geertz, "Ethos, World-View and the Analysis of Sacred Symbols," in *Man Makes Sense: A Reader in Modern Cultural Anthropology*, eds. Eugene A. Hammel and William S. Simmons (Boston, 1970), p. 326.

/7/   "Dans les sociétés ou la divination ne revêt pas, comme dans la nôtre, le caractère d'un phénomène marginal, voire aberrant, où elle constitue une procédure normale, régulière, souvent même obligatoire, la logique des systèmes oraculaires n'est pas plus étrangère à l'esprit du public que n'est contestable la fonction du devin. La rationalité divinatoire ne forme pas, dans ces civilisations, un secteur à part, une mentalité isolée, s'opposant aux modes de raisonnement qui règlent la pratique du droit, de l'administration, de la politique, de la médecine ou de la vie quotidienne; elle s'insère de façon cohérente dans l'ensemble de la pensée sociale, elle obéit dans ses démarches intellectuelles à des normes analogue, tout de même que le statu du devin apparaît très rigoureusement articulé, dans la hiérarchie des fonctions, sur ceux des autres agents sociaux responsable de la vie du groupe. Sans cette double intégration de l'intelligence divinatoire dans la mentalité commune et des fonctions du devin dans l'organisation sociale, la divination serait incapable de remplir le rôle que lui ont reconnu les anthropologues de l'école fonctionaliste." J. P. Vernant, "Parole et signes muets," in *Divination et Rationalité*, ed. J. P. Vernant (Paris, 1974), p. 10.

/8/   I have estimated (*Sources of Shang History*, p. 89) that some fifty man-hours a day were devoted to scapulimancy and pyromancy during the reigns of the last eight or nine Shang kings.

/9/   There is not a great deal of contemporary evidence for Shang rituals directly concerned with the divinatory act. It is hardly likely, however, that the Shang contacted the supernatural without ritual. It is believed that the scapulas and turtle shells were ritually prepared before use, and that certain ritual rules governed the burning of the cracks (*Sources of Shang History*, pp. 12–17, 26, n. 120). That divination occurred in the ancestral temple (*Nan-pei*, "Ming" 729; *Chin-chang* 120; *Ch'ien-pien* 8.15.1 [all S270.2]; *Yi-ts'un* 131; *Hou-pien* 2.42.15; *T'ieh-yi* 1.10; *Ts'uo-pien* 12; *Chih-hsu* 64 [all S270.3]. Here and below, S refers to Shima Kunio 島邦男 , *Inkyo bokuji sōrui* 殷墟卜辭類 [2d rev. ed.,

Tokyo, 1971]; the full bibliographic references for the abbreviations by which I cite oracle-bone inscriptions may be found in *Sources of Shang History* pp. 229–31), and that the verb for "divining" was written 貞 or 鼎 , a simplified picture of a cauldron (modern *ting* 鼎), are further reasons for thinking that rituals were involved. Such a hypothesis is confirmed by the elaborate descriptions of plastromantic ritual in later texts such as *Chou-li* and *Li-chi*.

/10/   On the way in which ritual confers an aura of factuality on religious conceptions, see Clifford Geertz, "Religion as a Cultural System," in *Anthropological Approaches to the Study of Religion*, ed. Michael Banton (London, 1966), pp. 24–26.

/11/   Ambiguous prognostications, the diviner's defense against error, are common to many forms of divination around the world. "It was a notorious fact in antiquity that Apollo's oracular responses [at Delphi] were crooked and ambiguous" (H. W. Parke and D. E. W. Wormell, *The Delphic Oracle: Vol. 1: The History* [Oxford, 1956], p. 40). It was a recurrent theme in stories about prophecy in classical Greece "that the recipient should misunderstand the message and end in trouble, but that when the full facts are known the prophet's foreknowledge should be vindicated" (Antony Andrewes, *The Greeks* [London, 1967], pp. 242–43). There was no such tradition in China.

/12/   For an introduction to Tung Tso-pin's five periods, see *Sources of Shang History*, pp. 92–93, 203.

/13/   *Ping-pien* 247.1–2 (S140.1).

/14/   In this regard, the Shang diviners were by no means unique. Consider Alfred Guillaume, *Prophecy and Divination Among the Hebrews and Other Semites* (London, 1938), p. 47, on Sumero-Babylonian divination: "In all these oracles it is the schemes of men that are the concern of the gods, not the will of God that is the concern of man."

/15/   I have argued this view in my paper, "Shang Divination and Shang Metaphysics (With an Excursion into the Neolithic)," Berkeley, March 1973, mimeographed, pp. 13–19.

/16/   For the Hebrew case, see Guillaume, *Prophecy and Divination*, pp. 107 ff.

/17/   See my paper, "Legitimation in Shang China," Conference on Legitimation of Chinese Imperial Regimes, 15–24 June 1975, Asilomar, California, mimeographed, pp. 9–24.

/18/   The essential characteristics of "display inscriptions" are: (i) bold, large calligraphy; (ii) the prognostication and verification are written as a single, continuous unit, and are usually placed immediately next to the charge; (iii) the verification, freqently detailed, confirms the accuracy of the prognostication. *Ping-Pien* 57.1; 247; *Ching-hua* 2, are typical examples; the last two are translated in *Sources of Shang History*, fig. 12, and p. 44.

/19/   This, at least, is the picture presented by the record. Whether the whole record was fabricated after the event, so that the forecast of misfortune was not recorded till after the misfortune occurred, need not concern us here. For an introduction to the issues involved, see *Sources of Shang History*, p. 46, n. 90.

/20/   On this function of the particle *ch'i*, see Paul L-M Serruys, "Studies in the Language of the Shang Oracle Inscriptions," *T'oung Pao* 60.1–3 (1974) 25. I offer "perhaps" only as a functional indicator; Serruys himself (p. 58) rejects it as a translation. There is also some evidence that desired alternatives were placed on the right side of the turtle shell, undesired and abbreviated alternatives on the left (see *Sources of Shang History*, p. 51, n. 124, p. 52, n. 130).

/21/   Again, the Shang were not unique in this view of the diviner's function. "The whole of the Semitic world was permeated with the belief that the solemn pronouncements of accredited persons—whether priests, prophets, diviners, or magicians—possessed an authority not only over the mind, but also over the course of events, so that what such men said must surely come to pass because they spoke in the name, and with the authority, of a supernatural power. Theirs was . . . the authority of the divine over the human; or, in reference to a more primitive stage, of what anthropologists call Mana" (Guillaume, *Prophecy and Divination*, p. 25).

/22/   The fact that, like the inscriptions, the *pu*-shaped stress cracks were, on occasion, themselves incised more deeply into the surface of the shell or bone (*Sources of Shang History*, p. 53, n. 135), is suggestive. The motive may have been related to Arthur Waley's observation that "an omen is regarded as in itself a momentary, evanescent thing. Like silver-prints, it requires 'fixing'. Otherwise it will refer only to the moment at which it was secured" ("The Book of Changes," *Bulletin of the Museum of Far Eastern Antiquities* 5 [1933] 136).

/23/   *Ch'ien-pien* 2.31.4 (S293.4); S165.1–168.2; *Hsü-pien* 4.17.8 (S170.1); *Ching-chin* 5132 and others at S343.1. For an account of the way divination forms had evolved by period V, see *Sources of Shang History*, p. 122. It may be noted that some of the changes in Shang divination practice were linked to the difference between the Old School of diviners (periods I, IIa, IV) and the New (periods IIIb, V), though the exact division between them, especially in III and IV is not firmly established (*Sources of Shang History*, pp. 32, n. 18, 203).

/24/   For an initial study of the noninterrogative nature of the divination charges, see my paper, "*Shih Cheng* 釋貞 : A New Hypothesis About the Nature of Shang Divination," Asian Studies on the Pacific Coast Conference, Monterey, California, 17 June 1972, mimeographed; see too, *Sources of Shang History*, p. 29, n. 7.

/25/   The phrase, presumably deriving from Frazer, is quoted by Geoffrey Parrinder, *Religion in Africa* (Baltimore, 1969), p. 64.

/26/   This may help explain the terseness of the verifications in the later periods (*Sources of Shang History*, p. 118), for if the recorded result always validated the forecast the content of the verification could virtually be taken for granted.

/27/   The same trend may be discerned in ancient Near Eastern divination. "Everywhere there is a tendency for the priest to succeed the prophet, and for the formal rites of religion to replace and supersede *mana*" (Guillaume, *Prophecy and Divination*, pp. 37–38).

/28/   The evidence for the three responses was presented in Keightley, "How the Cracks Were Read: The Existence of the Subcharge," a paper prepared for the Annual Meeting of the American Oriental Society, Toronto, 12 April 1978.

/29/   Robin Horton, "A Definition of Religion, and Its Uses," *Journal of the Royal Anthropological Institute* 90.2 (July-December 1960), 208–9, has remarked on the way that "explicit definition of a limited number of permissible responses" reduces uncertainties and dangers of communication, especially when a large status difference distinguishes the participants. He finds such stereotyping typical of religious contexts and opposed to the flexibility of nonreligious ones. The forms of Shang divination, reflecting perhaps the keenly felt status differences between man and the spirits, appear to have carried such stereotyping to extreme limits.

/30/   See n. 11 above.

/31/   For a typical case, see *Sources of Shang History*, pp. 78–79, n. 86.

/32/   *Ping-pien* 247.1 (S140.1).

/33/   *Sources of Shang History*, p. 36.

/34/   For a general account of the reduced scope of Shang divination by period V, see *Sources of Shang History*, pp. 122, 177–182.

/35/   E.g., *Ping-pien* 12–21 as discussed in *Sources of Shang History*, pp. 76–90.

/36/   The pyromantic theology implied in *Shang-shu*, "Chin-t'eng," may represent this transitional stage—the reluctance to trouble the spirits, the initiative taken by the Duke of Chou, the *do ut des* bargaining with the spirits, the argument couched in terms of how human abilities might please the spirits; the emphasis is still magical, rather than religious, but human virtues are given prominence. For a study of nascent Shang moral conceptions, see David S. Nivison, "Royal 'Virtue' in Shang Oracle Inscriptions," *Early China* 4 (1978–79) 52–55.

/37/   Keightley, "The Religious Commitment: Shang Theology and the Genesis of Chinese Political Culture," *History of Religions* 17.3 & 4 (February-May 1978): 211–25; "Public Work in Ancient China: A Study of Forced Labor in the Shang and Western Chou," Ph.D. dissertation, Columbia University, 1969, pp. 340–46.

/38/   For possible links between shamanism, conceptions of the soul, metaphysics, and ethics, see E. R. Dodds, *The Greeks and the Irrational* (Berkeley, 1963), esp. pp. 139–40.

/39/   On this point, see T'ang Chün-yi 唐君毅 , "Lun Chung-hsi che-hsüeh chung pen-t'i kuan-nien chih yi-chung pien-ch'ien pien-ch'ien 論中西哲學中本體觀念之一種變遷 ," in T'ang Chün-yi, *Chung-hsi che-hsüeh ssu-hsiang pi-chiao yen-chiu* 中西哲學思想比較研究 (Shanghai, 1947), pp. 123–146. See too, n. 38 above.

/40/   Particularly apposite is the Mo-tzu comment quoted in *Shih-chi*: "I have heard it said that he who looks into the water will see the form of his face, but he who looks at men will know fortune and misfortune." As Burton Watson notes, the terms *chi* and *hsiung* are "part of the terminology of the ancient art of divination" (*Ssu-ma Ch'ien: Grand Historian of China* [New York, 1958], p. 136).

/41/   A. Leo Oppenheim, *Ancient Mesopotamia: Portrait of a Dead Civilization* (Chicago, 1964), p. 210.

/42/   The evidence, which I hope to present in a forthcoming study, suggests that in most cases used oracle-bones were stored above ground until they occupied too much room; they were then dumped into a storage pit, either by themselves, or along with raw bone materials and other debris of Shang life. Ssu-ma Ch'ien had heard that the Hsia and Yin threw away their divining stalks and shells after use because they felt that stored plastrons were not spiritually efficacious (*pu-ling* 不靈) (Takigawa Kametarō 瀧川龜太郎 , *Shiki kaichū kōshō* 史記會注考證 [Tokyo, 1934], *ch.* 128, p. 3); I believe he was correctly informed.

/43/   Herbert Fingarette, *Confucius: The Secular As Sacred* (New York, 1972), pp. 18–36.

/44/   E.g., *Lun-yü*, 4.13; 10.1–2; *Chung-yung*, 20.2. (In these and the following citations to the Classics, I use the chapter and paragraph numbers of Legge's translations.)

/45/   E.g., *Lun-yü*, 9.28; 14.30. Cf. Fingarette, pp. 39, 43, 47.

/46/   The assertive self-confidence of early Chou religion is well demonstrated by the "Chou-sung" section of *Shih-ching*. There is no abasement before the powers. The major thrust is to proclaim the virtuous acts of the Chou, past and present, that have been, and should be, acceptable to the spirits.

/47/   Wing-tsit Chan, *A Source Book in Chinese Philosophy* (Princeton, 1963), pp. 40–41.

/48/   For the critical significance of this faith in *cheng-ming*, cf. Oppenheim, *Ancient Mesopotamia*, p. 231: "The fateful concept that reality should adjust to the requirements of a written corpus remains unknown to Mesopotamia—and probably to the entire ancient Near East. Only in a late and definitely peripheral development that sprang from the desire to create, for ideological reasons, a specific social context did Judaism succeed in creating such a pattern of behaviour."

/49/   W. Kohler, *The Place of Value in a World of Fact* (New York, 1938), p. 24.

/50/   *Lun-yü*, 7.29. Fingarette, pp. 3–4, argues that occasional comments in *Lun-yü* "seem to reveal a belief in magical powers of profound importance." The *chün-tzu* "simply wills the end in the proper ritual setting and with the proper ritual gesture and word; without further effort on his part, the deed is accomplished."

/51/   Chan, pp. 86–87.

/52/   Waley, p. 128, notes: "It is easy to see how the formula of folk-poetry (so often exemplified in the *Odes*) in which a series of statements concerning natural phenomena, trees, birds, etc., are correlated to a series of statements concerning a human situation grew up out of the omen formula." He is referring here to *Yi-ching* formulas, but the model for such correlations goes back to at least the divination charges of Shang.

/53/   These ideas about logic are taken from Donald M. Johnson, "Reasoning and Logic," in *International Encyclopedia of the Social Sciences*, ed. David L. Sills (New York, 1968), 13: 344.

/54/   *Lun-yü*, 2.5; Legge tr.

/55/   Modern scholars use the term "charge" (*ming-tz'u* 命 辭) to refer to the topic of the divination inscription. That the term *ming* was used in Chou plastromancy suggests that it may have been a Shang term, or that it at least reflects the spirit of Shang divination. For further details, see *Sources of Shang History*, p. 33, nn. 20, 21.

/56/   On the obligatory mood of many Shang charges, see *Sources of Shang History*, p. 66, n. 44.

/57/   Love of order is particularly strong in thinkers such as Hsün-tzu, Han Fei-Tzu, and Shang Yang. It may also be found in *Lun-yü*. "Confucius' philosophy is that of a feudal kingdom, and of a cosmic order in which everything has its properly ordained place; insubordination is perhaps the chief crime, according to the sage" (Herrlee Glessner Creel, "Was Confucius Agnostic?" *T'oung Pao* 29 [1932], 85).

/58/   See Jacques Gernet, *Daily Life in China on the Eve of the Mongol Invasion* (New York, 1962), p. 203; Mou Tsung-san 牟 宗三 , *Chung-kuo che-hsüeh ti t'e-chih* 中國哲學的特質 (Hong Kong, 1963).

/59/   These considerations have some bearing on the picture of Hsün-tzu presented by Henry Rosemont, Jr., "State and Society in the *Hsün-Tzu*: A Philosophical Commentary," *Monumenta Serica* 29 (1970–71): 38–78. I suspect that Hsün-tzu was predisposed to seek solutions to social problems in the *li* and in hierarchy because such solutions were inherently attractive. His conclusions about the economy and society of Warring States China were culturally determined.

/60/   Frederick W. Mote, *Intellectual Foundations of China* (New York, 1971), p. 71.

/61/   Hexagram 63, the image; *The I Ching or Book of Changes*, The Richard Wilhelm Translation rendered into English by Cary F. Baynes (Princeton, 1969), p. 245.

/62/   *Chung-yung*, 24, Legge translation. On the foreknowledge of the *chün-tzu*, cf. Moss Roberts, *"Li, Yi*, and *Jen* in the *Lun Yü*: Three Philosophical Definitions," *Journal of the American Oriental Society* 88 (1968), 765–66.

/63/   For example, I count eight forecasts of this type for the year, Chao 1, seven for the year, Chao 3. See, too, n. 40 above.

/64/   A particularly apposite example is provided in *Shih-chi*, when Chi Tzu forecasts the degenerating behavior of his relative, the last Shang ruler (Takigawa, *ch.* 38, p. 6; Edouard Chavannes, trans., *Les Mémoires historiques de Se-ma Ts'ien* [Paris, 1967], 4: pp. 216–17).

/65/   For a detailed study of this theme, see Sarah Allan, *The Heir and the Sage: A Structural Analysis of Ancient Chinese Dynastic Legend* (San Francisco, 1980).

/66/   Tu Erh-wei 杜而未, "Lu-chün chien-chiao pien-cheng 魯君僭郊辨證," *Kuo-li T'ai-wan ta-hsüeh k'ao-ku jen-lei-hsüeh k'an* 國立臺灣大學考古人類學刊 25–26 (November 1965), 25–26; Barry Burden Blakeley, "Regional Aspects of Chinese Socio-Political Development in the Spring and Autumn Period (722–464 B.C.: Clan Power in a Segmentary State," Ph.D. dissertation, University of Michigan, 1970, pp. 330, 331.

/67/   *Lun-yü*, 15.10.

/68/   Hans Bielenstein, "The Restoration of the Han Dynasty: Vol. 3: The People," *Bulletin of the Museum of Far Eastern Antiquities* 39 (1967): 36, 37.

/69/   John K. Shryock, *The Origin and Development of the State Cult of Confucius* (New York, 1966), p. 100.

/70/   Norman O. Brown, *Love's Body* (New York, 1966), p. 254. Brown is writing of psychoanalysis, which "began as a further advance of civilized (scientific) objectivity; to expose remnants of primitive participation, to eliminate them. . . . But the outcome of psychoanalysis is the discovery that magic and madness are everywhere, and dreams is what we are made of." The same insight, I suspect, applies to the secularization of societies.

The Speech of Prince Chin:
A Study of Early Chinese Cosmology

James A. Hart

The "Chou yü" 周語 ("Speeches of Chou") is the first three chapters of the *Kuo yü* 國語 ("Speeches of the States"). It consists of a series of speeches, usually by advisors to the Chou kings, arranged in chronological order, covering a period from approximately 1000 to 500 B.C. Each of the speeches deals with a particular, concrete problem of governmental policy, but the problem at hand always provides an opportunity for a generalized, philosophical discussion by the advisor making the speech.

The longest speech in the collection and one of the most interesting is supposed to have been made by Prince Chin to his father, King Ling, in 549 B.C. In that year there was a great flood which threatened to destroy the royal palace. The king wished to block off the flooding rivers to protect his palace, but his son urged him not to do so. He argued that his father should not look merely at the immediate problem of saving the palace, but that he should look at the situation from a larger perspective. First the prince approached the problem from the viewpoint of cosmology, explaining the pattern of nature to which men must conform. Then he encouraged his father to look at his position as Chou king from a historical perspective. Finally Prince Chin brought all the elements he had discussed together, showing that they all pointed in the same direction: that his father's proposed action would be wrong. The prince's speech extensively uses the rhetorical devise of parallelism in such a way that it strongly reinforces the argument being presented.

My treatment of the prince's speech will be arranged as follows:
1. Introduction—concerning problems of authorship and dating
2. Translation
3. Discussion of the cosmology presented in the speech
4. Discussion of the theory of history of the prince
5. Discussion of the relation between the ideas of the speech and the rhetorical techniques employed
6. A final note on the problem of dating

1. Introduction

To what extent can this section of the "Chou yu" be regarded as an authentic historical record? There is evidence that there was a flood, as it is mentioned in both the *Ch'un-ch'iu* 春秋 and the *Tso-chuan* 左傳 , although there is a one year discrepancy in dating (Duke Hsiang 相 , 24th year, 548 B.C.). The *Tso-chuan* also states that the king's wall was rebuilt in that winter by people from the state of Ch'i. However, there is no mention of Prince Chin or of any proposal to block the flooding rivers.

The records we do have of Prince Chin say nothing about this speech, and are not reliable history, in any case. In the *I chou shu* 儀周書 , for example, there is an account of a long conversation between Prince Chin at the age of fifteen and the blind music master K'uang of the state of Chin. The clever music master was amazed at the skill with which the prince discussed the ancient sage rulers, and was discomfited at the subtlety with which he criticized Chin's desire for expansion (at the expense, in part, of the royal domain). The Music master praised the prince, saying,

> "You will become the most esteemed person in the world!" The prince replied, "Music-master, why do you make fun of me? From the time to T'ai Hao (Fu Hsi) down to Yao, Shun, and Yü, there has never been a family which has regained the empire. If in this great affair I respond to the times and do not attack Heaven, then how can I succeed? Furthermore, I have heard that you know whether a person will have a long or short life. Tell me about my life."
>
> The Music-master replied, "Your voice is clear and expansive, so your complexion is red and white. A fiery complexion does not live long."
>
> The prince said, "It is so. After three years I shall ascend to become a guest at the place of God. You must be careful to say nothing, or calamity will befall you."
>
> The Music-master returned home, and before three years had passed, a messenger announcing the prince's death arrived. (IX, 64)

As in the speech in the "Chou yü," the prince is here an excellent speaker, and he is again pessimistic about the future of the Chou dynasty. Here the prince has the unusual ability to foresee his own early death.

An even more remarkable biography of Prince Chin appears in the *Lieh-hsien chuan*:

> Wang-tzu Ch'iao (the King's Son Ch'iao) was Chin, the oldest son of the Chou King Ling. He loved to play the *sheng* flute and

imitate the sound of the phoenix. He would wander about between the I and Lo Rivers. With the Taoist adept Fu-ch'iu Kung accompanying him, he ascended Sung-kao Mountain.

More than thirty years later, when they were looking for him in the mountains, he saw Huan Liang and said, "Tell my family to wait for me on the seventh day of the seventh month on the peak of white crane. He stopped on the summit of the mountain, and they could see him from a distance, but were unable to reach him. Raising his hand, he bade farewell to his contemporaries. Several days later he was gone. They built shrines to him at the foot of Kou-shih Mountain and on the peak of Sung-kao Mountain.

Marvellous was Wang Tzu!
With wandering spirit and vigorous *ch'i*;
He played the flute and sang by the I and Lo,
Fu Ch'iu, with responsive mind,
Took his hand and led him aloft.
Waving his whip from a green cliff,
On borrowed wings he went on alone. (I, 28)

As Wang-tzu Ch'iao, Prince Chin became a Taoist immortal, to reappear many times over the centuries in Chinese literature and art. He is mentioned as early as the *Ch'u Tz'u* 楚辭 and is the subject of several biographies in the *Tao-tsang* 道藏 . However, I have found nothing in the Wang-tzu Ch'iao legend that helps to further identify Prince Chin, the purported maker of the speech.

The speech is supposed to have been made in the middle of the sixth century B.C., around the time of the birth of Confucius. If it could be reliably attributed to that period, then it would be a very significant contribution to early Chinese philosophy. However, it is very doubtful that the speech does date from this early time. The speeches of the "Chou-yü" cover a period of some five centuries, and yet they show a striking uniformity in language and style. There is also a remarkable consistency in the ideas being presented. It is true that there are a number of different topics covered, but when the same topic appears in different speeches, the advice offered by different speakers is never contradictory. Furthermore, while a number of different subjects are covered, the advice in the different speeches fits together to form a coherent system. This uniformity indicates that all of the speeches are probably in large part the work of the editor (or author) who put them in their final form.

Who was this editor, and when did he live? We do not know. The *Kuo-yü* is first mentioned by Ssu-ma Ch'ien, and in his autobiography and his letter to Jen-an, he attributes it to Tso Ch'iu-ming, a shadowy figure about whom we know little. As Tso Ch'iu-ming is also supposed to have been the author of the *Tso-chuan*, these two works, the *Tso-chuan* and *Kuo-yü*, have been traditionally associated as works by a single author. However, there were also scholars who have questioned this assumption (including

Ssu-ma Kuang of the Sung and Ts'ui Shu of the Ch'ing dynasty).

The *Tso-chuan* has been a subject of great controversy ever since the first century B.C., when Liu Hsin proposed that it be accepted as an official commentary for the classic, the *Ch'un-ch'iu*. This controversy was greatly heightened in the nineteenth century when first Liu Feng-lu and then, more importantly, K'ang Yu-wei contended that it was a forgery. K'ang noted the fact that Ssu-ma Ch'ien mentioned by name the *Kuoyü*, but never the *Tso-chuan*. K'ang then went on to argue that Liu Hsin took the original *Kuo-yü* and made it into a faked commentary on the *Ch'un-ch'iu*, and gave it the name, the *Tso-chuan*. Then from fragments available in the imperial archives he put together a new *Kuo-yü*, so that suspicion would not be aroused by the loss of a book whose title was known. Thus, according to K'ang Yu-wei, the *Tso-chuan* and *Kuo-yü* which we have are both forgeries by Liu Hsin./1/

K'ang's argument has been effectively refuted, especially by Bernard Karlgren and Henri Maspero. Karlgren argued that the *Tso-chuan* has a consistent and unique grammatical system which could not have been invented by a Han forger, but which must have represented a dialect of the Warring States period. He also found that the *Kuo-yü* is very close, but not identical in language, to the *Tso-chuan*, indicating that the two works represent essentially the same dialect (and probably the same school) but were by different authors./2/

Maspero rejected the theory that Liu Hsin proposed the *Tso-chuan* because it would help the usurper Wang Mang. Maspero showed that Liu began to champion the *Tso-chuan* long before Wang came to power, at a time when Wang was in disgrace, and because of the political situation could not have had expectations at all of coming to power. Maspero also showed that the astronomical and calendrical calculations of the *Tso-chuan* were based on observations which could not have been made either in the Ch'un-ch'iu times or during the Han dynasty, but could only have been made during the Warring States period./3/ More recent works on this subject tend to substantiate Karlgren and Maspero, but without greatly increasing the scope of their arguments.

In my opinion these arguments taken together effectively discredit K'ang Yu-wei's theory that the *Tso-chuan* and *Kuo-yü* were forgeries by Liu Hsin. Thus, we can reasonably assume that the speech of Prince Chin was written during the Warring States period, although exactly when and by whom is unknown.

The text I use is that in the *Ssu-pu pei-yao, Kuo-yü* III, 5a–9a. This is one of two principal texts in circulation today. The other, which is in the *Ssu-pu ts'ung-k'an*, has some differences, but they are few and minor. A compilation and study of variants found in these texts and other surviving fragments has been made by Chang I-jen, *Kuo-yü chiao-cheng* 國語校正. Pages 100–108 refer to the prince's speech.

Almost all editions of the text include the commentary of Wei Chao 韋昭 (197–278), which I have found very helpful.

2. Translation

In the twenty-second year of King Ling 靈王 (549 B.C.), the Ku 谷 and Lo 洛 Rivers were on the rampage, threatening to destroy the royal palace./4/ The king wished to dam up the rivers, but Crown Prince Chin admonished him, saying, "This must not be done. I have heard that those who ruled the people in ancient times did not tear down the mountains, nor did they raise the marshes, nor did they obstruct the rivers, nor did they drain the swamps. For mountains are accumulations of earth, and marshes are gathering places for creatures. Rivers are channels for energy, and swamps are concentrations of water.

"When heaven and earth became complete, they had accumulated the high (mountains) and gathered creatures in the low (marshes). They had cut through rivers and valleys, to channel their energy, and had dammed and dyked stagnant and low-lying water, to concentrate their fertility. For this reason the accumulations (mountains) do not crumble and collapse, and so creatures have (marshes) in which to gather. Energy is not sunken and congealed, nor is it scattered and dissipated./5/ Through this the people, when living, have wealth and useful things, and when dead have places to be buried. When this situation prevails, there will not be the sorrows of early death, confusion, pestilence, or plague, nor will there be the distresses of famine, cold, scarcity, and want. Therefore those of high rank and low can strengthen each other and thereby await the unforeseen. It was just this to which the sage kings of old were attentive.

"Formerly, Kung Kung 共工 abandoned this Way, enjoying himself with excessive pleasures./6/ Losing himself in dissipation, he wanted to dam the hundred rivers, tear down the high and fill up the low, thereby injuring the whole world. August Heaven did not bless him, nor did the people assist him. Calamity and disorder rose simultaneously, and Kung Kung was thereby destroyed.

"During the rule of Yü 禹 (Shun), there was Kun, Earl of Ch'ung 崇伯鯀 who indulged his licentious heart and followed approvingly the excesses of Kung Kung./7/ Yao 堯 therefore killed him on Mount Yü.

"His descendant Yü 禹 , remembering that his predecessor had violated the measures, reformed the rules and standards. He observed the patterns of heaven and noted the distinguishing characteristics of earth. He followed according to category the hundred principles, and used them as models for the people and as measures for the flocks of living things. The collateral descendants of Kung Kung, the Ministers of the

Four Peaks 四嶽 , assisted him./8/ They raised the high, lowered the low, dredged the rivers to channel the congealed, and concentrated the (swamp) waters to make abundant the living things. They elevated to lofty heights the nine mountains, and dredged to make free flowing the nine rivers. They banked and dammed the nine swamps, and made richly flourishing the nine marshes. They caused to flow and rush forth the nine springs. They made inhabitable and settled the nine secluded areas. They united in communication all within the four seas. Therefore heaven had no hidden yin, and earth had no scattered yang. There were no waters in which the energy was sunken, and no calamitous wild fires. There were no spirits with divisive conduct, and there were no people with licentious hearts. There were no seasons which violated the proper sequence, and no creatures which harmed living things./9/ They followed as pattern the accomplishments of Yü, and drew from them a guiding model. There was nothing which was not an admirable accomplishment, so that they were able to satisfy the heart of God.

"August Heaven rejoiced in (Yü) and rewarded him with the world. Yü was given the clan Ssu 姒 , and his family called Holders of Hsia 夏 . This signified that he could with admirable blessings give abundant wealth to living things./10/ The Ministers of the Four Peaks were given a state and it was decreed that they would be chief among the feudal lords. They were given the clan Chiang 姜 , and their family was called Holders of Lü 呂 . This signified that they could act as the legs, arms, heart, and backbone for Yü in nourishing creatures and making the people prosperous./11/

"How could it be that this one king and four lords received so many blessings, as they were descendants of destroyed kings?/12/ It was only because they could order and establish admirable justice that they have had descendants existing into later generations, who have preserved the sacrifices and not neglected their rules. Even though the holders of Hsia state has declined, Ch'i 杞 and Tseng 鄫 still exist./13/ Even though Shen 申 and Lü 呂 have declined, Ch'i 齊 and Hsü 許 still exist./14/

"It was only because they had admirable accomplishments that they were granted clans and received sacrifices which extended throughout the world. And as to (their descendants) who lost (their high positions), they certainly had evil, licentious hearts, which separated them (from their ancestors). Therefore they lost their families and clans, and were not saved from untimely death. Their family lines were cut off with no one to lead the sacrifices, or they sank and became servants and grooms.

"Those who were destroyed—how was it that they were without blessings, for they were all descendants of the Yellow Emperor or Yen Ti 火帝 ./15/ It is just because they did not follow the measures of heaven and earth, did not accord with the order of the four seasons, did not take as measure the just order of the people and spirits, did not take as models the principles of living creatures: it was for these things that they were destroyed, and had no posterity, so that today their sacrifices are not performed.

"But as for (their descendants) who succeeded, they certainly had hearts of honesty and trustworthiness that separated them from (their ancestors). They took as measures heaven and earth, and accorded with the movements of the seasons. They harmonized with the people and spirits, and took as models the principles of things. Therefore, exalted and shining, they died honorably; illustrious and enduring shone their brilliance. They were granted their clans and families, and they added to them an ilustrious name.

"If one investigates the teaching inherited from the former kings and examines their codes of regulations and their criminal laws, and then observes who succeeded and who was destroyed, then all can be known. Those who succeeded certainly had achievements such as Hsia and Lü./16/ Those who were destroyed certainly had failures such as Kung Kung and Kun. Now that today we hold the government, are there not things which we must avoid? And yet still you would disturb the spirits of the two rivers, forcing them even to struggle with each other, thereby endangering the King's palace. For the King himself to embellish it—is this not something that must not be done?

"People have a saying which goes, 'Do not pass the gate of a disorderly man.' Again, they say, 'Assist at a feast, and get to taste. Assist at a fight, and get hurt.' Again, they say, 'If calamity is not relished, then it cannot become a calamity.' The *Book of Poetry* says, 'The four stallions are vigorous; the falcon and reptile banners are fluttering. Disorders arise, and are not quieted; there is no state which is not being destroyed.'/17/ It also says, 'The people covet disorder: they find contentment in acting as bitter poison.'/18/ If one sees disorder and is not fearful, its destructive consequences will certainly be numerous, and in the end your embellishment will be displayed. When the common people have cause to be resentful and disorderly, even they cannot be checked, so how much more so in the case of the spirits! If the King is going to obstuct the fighting rivers in order to embellish his palace, this is to embellish disorder and assist fighting. Would this not then be displaying calamity and welcoming injury?

"From the time that our ancestors, Kings Li, Hsüan, Yu and P'ing 厉 宣 幽 平 coveted the calamities from heaven, down to today, these calamities have not ceased. If we in addition display them, I am afraid that they will reach down to our descendants, and that the royal house will become increasingly humbled.

"How is this? From the time of Hou Chi 後稷, they quieted disorder, but not until the time of Kings Wen, Wu, Ch'eng, and K'ang 文 武 成 康 could they bring peace to the people. From the time when

Hou Chi first laid the foundation for making the people tranquil, it was fifteen generations later before King Wen began to bring them order. Only after eighteen generations was King K'ang able to bring them peace.

"Their difficulties were similar to this. It was fourteen generations ago that King Li first changed the laws. Only after fifteen kings had laid the foundations for virtue did they first bring order. When fifteen kings have laid the foundation for calamity, might it not be beyond saving?/19/ I am from morning to night alarmed and fearful, and say, 'what virtue may we cultivate, that we may to a small degree brighten the royal house, in order to welcome the blessings from heaven?' If the king in addition displays and assists calamity and disorder, how can we bear it?

"Does the king not also have a mirror in the kings of the Li and Miao 黎 苗, and later in the last rulers of Hsia and Shang 夏 商? Above they did not follow the patterns of heaven; below they did not take as model the earth; in the middle they did not harmonize with the people; on all sides they did not accord with the seasons, and they did not provide for the spirits. Thus they discarded and abandoned these five principles. For this reason men destroyed their ancestral temples and burned their sacrificial vessels, and their sons and grandsons became servants. They were not at peace with the common people, and they did not observe the principles of ancient wisdom and excellent virtue.

"But if one takes as principle these five,/20/ and one receives abundant prosperity from heaven and is blessed with the people's loyal efforts, then one's descendants will be wealthy and numerous, and one's honorable reputation will not be forgotten. All of this is what you, the Son of Heaven, knows.

"Of the descendants of those whom Heaven has honored, some are working in the fields—this is because they wished to bring disorder among the people. And of the people from the fields, some are at the Altars of Earth and Millet—this is because they wished to make the people tranquil. There is no case which is different from this. The *Book of Poetry* says, 'The mirror of Yin is not distant; it is in the generations of the Hsia lords.'/21/ What then is the use of embellishing the palace? Is this not thereby seeking disorder? If one measures it by heaven and the spirits, then it is inauspicious. If one compares it with the earth and things, then it is not just. If one arrays it with the principles of the people, then it is not humane. If one compares it to the movements of the seasons, then it does not accord. If one considers it with the teachings of old, then it is not correct. If one observes it in the *Book of Poetry* and *Book of Documents*, and in the exemplary sayings of the common people, then in each case it is the action of lost kings. If one discusses it from above or below, then there is nothing with which it is compared or measured. You, the King, should consider it.

"As to this affair, on the large scale it does not follow the patterns, and on the small scale it does not follow the markings./22/ Above it violates the form of heaven, and below it violates the nature of earth. In the middle it violates the principles of the people, and on all sides it violates the movements of the seasons. Anyone who would initiate such a policy is certainly unrestrained. To initiate such a policy and to pursue it without restraint is the Way of Harm."

The King in the end blocked up the rivers. Later, during the reign of King Ching 景王 , the King had many favorites, and at this point disorder first arose./23/ When King Ching died, the royal house was in great disorder. During the time of King Ting 定王 the royal house was consequently humbled./24/

3. Cosmology

The Prince began his argument with the statement: "I have heard that those who ruled the people in ancient times did not tear down the mountains, nor did they raise the marshes, nor did they obstruct the rivers, nor did they drain the swamps." We will begin by discussing (in the following order) the rivers, swamps, marshes, and mountains, to see why the prince did not want them disturbed.

Rivers. Floods were a perennial problem in China. If the flow of the silt laden rivers of north China were impeded, then they would spread over the countryside, wreaking havoc among the inhabitants. Prince Chin warned his father against blocking the rivers because they are "channels for the energy" of nature. The word "channel" is *tao* 道 "to lead, guide." It is glossed in Wei Chao's commentary as *ta* 達 "to teach, penetrate," here used in a causative sense. Thus the rivers are channels which guide or lead the energy, causing it not to be obstructed, but to flow through. The word "energy" is *ch'i* 氣 . It is capable of many different translations and meanings, all related. In nature, it can mean air, vapor, or steam, but it does not carry these meanings here. It is rather taken analogically from human physiology as understood by the Chinese. In human terms it fundamentally means "breath," but it was understood, probably as a result of meditation practices, to circulate not only in and out of the lungs, but also throughout the body, as a kind of vital energy. Wing-sit Chan defines *ch'i* in this ancient period as "the psychophysiological power associated with blood and breath."/25/ It is in this sense that it is used in this speech. This analogical use of *ch'i* is specifically stated in the *Kuan tzu*: "Water is the blood and *ch'i* of the earth, circulating in a manner similar to the blood vessels" (XIV, 39).

In human terms it is very important that the circulation of *ch'i* be unimpeded. In the *Kuan-tzu* it is stated, "If the *ch'i* circulates, then there is life." (XVI, 49). Later in the same section is this warning: "If you overeat, you must exercise energetically. If you do not exercise energetically, then the *ch'i* will not circulate to the four extremities." The results of a lack of flow are given specifically in the *Lü-shih ch'un-ch'iu.*

> If the body is not exercised, then the essence will not flow. If the essence does not flow, then the *ch'i* will stagnate. If the stagnation is concentrated in the head, then it will become swollen and one will develop insanity. If it is concentrated in the ears, then (their passages) will be constricted and one will develop deafness. If it is concentrated in the eyes, then they will become sore and one will develop blindness. If it is concentrated in the nose then one will catch cold and become stopped up. If it is concentrated in the intestines, then they will become distended and engorged. If it is concentrated in the feet, then they will become paralytic and one will stumble. (III, 2)

Thus the analogy becomes meaningful. In both the human body and the earth as a whole, the flow of energy must be unimpeded. Otherwise it will become diverted in harmful ways.

This analogy is also used in a third area, in relation to human society. In another speech in the "Chou yü," the Duke of Shao warned King Li against his policy of prohibiting criticisms of his government. The Duke said, "To block up the people's mouths is even more extreme than blocking up rivers. If a river is obstructed and breaks through, then the injury to people will necessarily be great. The poeple are also like this. Therefore, those who control the rivers dredge them out, causing them to flow. Those who control the people open channels of communication, causing them to speak" (I, 3).

He then outlined the methods of providing each social group within the country with an appropriate channel of communication.

Here we have a pattern which holds on three levels. The flow of energy must not be obstructed on any level. Within the body obstruction will cause physical or mental disorder. On the social level it will lead to resentment and rebellion. In nature it will cause floods.

Swamps. However, while the flow of waters must remain unimpeded, according to Prince Chin, it must not be carried to excess. Just as the wise rulers of old did not obstruct the rivers, so also they did not drain the swamps. As rivers are to channel the water, causing the energy to flow, swamps are to gather together the waters and thereby compound their fertility. When with both rivers and swamps the proper balance is reached, then the "energy is not sunken and congealed, nor is it scattered and dissipated."

This is seen in terms of yin and yang. The rivers are the yang, the active force, which is necessary, but which in excess will lead to scattering and dissipation. The swamps are yin, passive, likewise necessary, but which in excess will lead to the energy's being sunken and congealed. This notion of the balancing of the active yang and the passive yin is expressed once in this speech as follows: when the world is functioning in proper order, then, "Heaven has no hidden yin, and earth has no scattered yang." In another speech in the "Chou yü," it is stated that when all the different musical instruments are played together in proper balance, then, "ch'i will have no congealed yin nor any scattered yang" (III, 6).

The practical application of the theory in relation to rivers can be seen thus: when rivers are flowing, the active, yang, movement of energy is apparent. If the flow is stopped, then the energy has apparently disappeared, as the waters become still. However, in fact, the energy has only become "sunken and congealed," and still exists in potential form, for if the obstruction is broken through, then the energy will appear again, in perhaps a violent fashion. The greater the obstruction, the greater the potential for destuctive force when it does erupt. Thus the flow must not be impeded. The proper compounding of the energy of water does take place in swamps, and it may be inferred that the speaker sees there the transfer of energy from the motion of the rivers to the fertility of the swamps, though this point is never explicitly stated.

Marshes. The word "marsh," *sou* 藪 , as distinct from "swamp," *tse* 澤 , should be explained. It has two connotations. First, Wei defines marsh as a "swamp without water." That is, it is a low, flat area that is humid, but without standing water. Secondly, because of these geographic characteristics, a marsh is an area of thick vegetation where wild animals find refuge. This meaning of marsh is seen in a poem in the *Shih Ching*, in which a marsh is set on fire during the dry winter months, enabling the hero to go in and kill a tiger (Ode 78). In the *Shang Shu* the evil King Chou is condemned as the "leader of the fugitives of the world who congregate around him like (fish) in a deep pool or (animals) in a marsh."/26/ The prince believed that the marshes should be preserved as refuges for the flocks of creatures. When heaven and earth are complete, there is the "gathering of creatures in the low (marshes)." He also stated that when the sage Yü discerned the "hundred principles" of heaven and earth, he used them as "measures for the flocks of living things." Wei explains "used as measures" to mean "did not injure or harm." I gather this to mean that Yü used the principles to establish limits to prevent excessive human encroachment against the domains of the wild animals.

Mountains. The prince warns that the mountains must not be torn down. He does not say specifically why they should not be torn down, except in their relation to the marshes. "The accumulations (mountains) do not crumble and collapse, and so creatures have (marshes) in which to gather." That is, if the mountains were torn down, this would necessarily

raise the low areas, thus making them drier, and unsuitable for dense
vegetation. Here there is expressed another yin-yang relationship. Just as
there is a close connection between the active rivers (yang) and the still
swamps (yin), so also there is a bond of mutual dependence between the
high mountains (yang) and the low marshes (yin).

After stating that these areas must be preserved, the prince describes
the good results that can be obtained: the people will be free of hunger,
cold, and plague, and the high and low can live together in harmony. Then
the prince appealed to history to support his case: Kung Kung violated
these natural regions, thereby bringing on himself natural calamity and
popular rebellion. Kun followed Kung Kung's example, and was put to
death by the emperor. But Kun's son Yü resolved to make up for his
father's failures, and it is from Yü that we can learn the proper way of
dealing with nature.

According to the prince, Yü "observed the patterns of heaven and
noted the distinguishing characteristics of earth. He followed according
to category the hundred principles, using them as models for the people
as measures for the flocks of living things." This is a key section of the
prince's speech, as it outlines succinctly his approach to understanding
and utilizing the world of nature. However, because of the brevity of the
passage I have had great difficulty in arriving at a translation and in
understanding the implications of the statement. As the passage is both
important and difficult I have below analyzed some of the key terms,
hoping thereby to justify my translation and to make clear some of the
implications, as I understand them.

The phrase "observed the patterns of heaven" is *hsiang t'ien*
象天 in Chinese./27/ The word *hsiang* has a variety of meanings:
elephant, ivory, image, imitate, star, omen. Han Fei-tzu explained the
relation between elephant and image as follows. "Men seldom see a live
elephant *(hsiang)* 象 . But when they obtain the bones of a dead elephant, they follow their pattern to visualize the living. That which
enables people to visualize in their minds is called an image *(hsiang)*"
(XX, 8). Whether this etymology is correct is unknown, but it is useful in
showing us how it was understood at the time. An image, then, is a pattern like the bones of an elephant which enables us through inference to
visualize that which we cannot actually see. The meaning of *hsiang* as a
star is related to this. The stars are images from which we can infer, for
example, the pattern of the changes of the seasons. The "Ta chuan"
states that in ancient times Pao Hsi observed the "images in heaven" as
well as other phenomena. Based on these observations he invented the
eight trigrams, which are supposed to embody the basic aspects of cosmic change (II, 2). Here, again, the stars *(hsiang)* suggest the cosmic
pattern, just as the bones of an elephant *(hsiang)* suggest its living form.

In the phrase at hand the wording is not *t'ien hsiang*, "images of heaven," but rather *hsiang t'ien*, "to imitate heaven." What does this mean? According to Tung Chung-shu, man is microcosm, imitating the universe in his physical being. "Above (men) have eyes and ears which are clear of hearing and clear of sight, and so are images of the sun and moon" (XIII, 56). This comparison holds good because the word "clear sighted," *ming* 明 , is composed of the characters for sun and moon and also means "bright." Thus the eyes imitate the sun and moon in being *ming*.

However, we "imitate heaven" in another way. The "Canon of Yao" in the *Shu ching* states that "(Yao) commanded the (brothers) Hse and Ho, in reverent accordance with the august heavens, to compute and delineate (*hsiang*) the sun, moon and stars, and the celestial markers, and so to deliver respectfully the seasons to be observed by the people." This translation is Needham's./28/ Legge/29/ and Karlgren/30/ also translated *hsiang* as "delineate" in this passage. Here, then, *hsiang* means not just to imitate, but to discern the pattern, as with the bones of the elephant. Hence, I have translated *hsiang t'ien* as "observed the patterns of heaven."

The next phrase *wu ti* 物地 I have translated "noted the distinguishing characteristics of earth." *Wu* usually means "thing." It also is used less frequently to mean "distinguishing characteristic or quality of a thing," as in the *Shih Ching*, "We matched *according to quality* (wu) 物 the four black horses" (Ode 177). With the same basic meaning it becomes a verb in this sentence from the Chou Li, "*According to quality they sorted* (*wu*) the horses and gave them out" (IV, 5). In another example in the *Tso chuan*, when preparations were being made to wall the capital, the different calculations included the following; "He *determined the distinguishing characteristics* (*wu*) of the lay of the land" (Chao, 32). In the prince's speech *wu ti* is similar in meaning: "He noted the distinguishing characteristics of earth."

After observing heaven and earth, Yü followed according to category the "one hundred principles," *pi lei pai tse* 比类百則 . The word "follow," *pi* 比 , also means compare, and so the phrase could also be translated "he compared and categorized the one hundred principles." I originally favored this translation, but I now think that it implies a detached objectivity on the part of the observer, which (although an ideal of Western science) was not characteristic of ancient Chinese thought. According to Wei's commentary, *lei* 类 , "category," is like *hsiang* 象 , "imitate," that is, just as nature has its categories such as yin and yang, so man should imitate such categories appropriately in his own conduct. Illustrations of how man follows the categories of nature will be given below. Here I simply want to suggest that *pi* indicates that man follows nature, and *lei* indicates that the categories of nature have their corresponding categories in human conduct.

What Yü "followed according to category" were the "one hundred principles," *pai tse* 百則 . "One hundred" is of course simply a general term, implying that Yü observed a great many different aspects of the world. Of these "hundred principles" the prince referred to the following in his speech: "principles of things," "principles of living things," "principles of the people," "principles of ancient wisdom and excellent virtue," and the "five principles," referring to heaven, earth, the people, the seasons, and the spirits. It is interesting to note that the prince never mentions any general "principles of nature." This bears out Needham's contention that *tse* 則 means "rules applicable to parts of wholes"/31/ and not "laws of Nature in the Newtonian sense."/32/ In the prince's speech there is implicit a general order of nature, but there is never an effort to abstract any general scientific laws.

The next statement shows more fully how Yü followed the hundred principles. He "used them as models for the people and as measures for the flocks of living things." The last part of this statement was discussed briefly above in the section of marshes, and will not be discussed here. Here we will focus on the term *i* 義 , which I have translated "used as models." The word *i* commonly means "good manners" or "rules of decorum." Twice in the *Shih Ching* it is used to mean "model (of correct behavior)" as in Ode 235.7 "Take King Wen as model."/33/ In the *Mo tzu* one chapter is entitled "Model," *fa i* 法義 (I, 4). In that chapter we are encouraged to take heaven, rather than any human, as our model. The idea of *i* as a cosmic model is found in the "Ta chuan," where the "two models," *liang i* 兩義 , refer to the unbroken yang line and the broken yin line. In Chang Heng's treatise on astronomy *Ling Hsien* (first century A.D.)/34/ the phrase "two *i*" refers to the sun and moon, which are concrete representations of the forces referred to in the "Ta chuan." Finally, for Chang Heng and other Han astronomers *i* was used to designate certain astronomical instruments such as armillary spheres, instruments whose bands of concentric rings served as models for the paths of the stars.

At first glance the two meanings of *i*, "good manners" and "scientific instrument," seem utterly unrelated. When we consider that *i* can be "*model* of correct behavior," and as scientific instrument *i* first meant "*model* of the heavens," then the meanings are a little less distant. When we remember that for many early Chinese thinkers there was an intimate connection between the pattern of nature and the correct pattern for human behavior, then it is easy to see how one word could be used as "model of the heavens" on the one hand and "model of human conduct" on the other.

How are the patterns of heaven used as a model for human affairs? The prince does not explain this, but in other speeches in the "Chou yü" there are three different ways mentioned which may be used to guide us. First, the movements of the stars serve as the most reliable guide to the changing of the seasons. In the following passage the changes of the stars are correlated with the changes in phenomena on earth, which in turn require corresponding changes in the occupations of men.

> When the constellation Ch'en Chüeh appears, the rain stops. When the constellation T'ien Ken appears, the rivers dry up. When the constellation Pen appears, the plants shed their leaves. When the constellation Ssu appears, frost falls. When the constellation Huo appears, the clear winds warn of cold.
>
> Thus the teachings of the former kings said, "When the rains stop, clear the roads. when the rivers dry up, complete the bridges. When the plants shed their leaves, store the harvest. When the frost falls, get ready the fur garments. When the clear wind comes, repair the inner and outer defense walls and the palaces and buildings." (II, 7)

Secondly, besides assisting us with the calendar, stars can be used in ways which now come under the heading of astrology. In one speech in the "Chou yü" King Wu is said to have picked a particularly auspicious time to attack Shang because the locations of the sun, moon, Jupiter, Mercury, and the point of the next conjunction of the sun and moon all indicated a favorable condition for the Chou house (III, 7).

Finally, the stars serve as a guide for human morality. In one speech we find the statement, "If a person follows the pattern of heaven, he can be respectful" (II, 2). Respect is a key virtue in Confucian morality, which stresses the proper relationship between superior and inferior. But how do the stars teach respect? This is explained by piecing together statements from several early texts. For example, in the *Lun yü* Confucius said, "If one conducts government through virtue, he may be compared with the North Star, which remains in its place while the multitude of stars pay homage to it" (II, 2). Thus, the one star, like the emperor, serves as the stable pivot around which the heavenly bodies revolve, at differing and even changing rates, but all according to a discernible pattern. This idea is expanded and applied to the family in the *Tso-chuan*. "We establish father and son, older and younger brothers, aunt and sister, first and second cousins, bridegroom and bride, marriage connections and in-laws to imitate (*hsiang*) the heavenly bodies" (Chao 25). Tu Yü's commentary explains, "The six relationships are in harmony, thereby serving the stern father, just as their myriad stars pay homage to the polar star."

Thus, the orderly pattern of the stars serves as a pattern for proper social relationships, leading Hsün-tzu to say, "It is through (ritual) that heaven and earth unite, that the sun and moonshine, that the four seasons have their sequence that the stars move, that the rivers flow, and that the ten thousand things prosper" ("*Li lun*").

Thus, while the prince does not tell us how we use the patterns of heaven as a guide for human conduct, other speeches in the "Chou yü," as well as other early Chinese texts, show us several ways in which human affairs should be guided by the stars.

The phenomena of earth also serve as our guide. As indicators of the changing of the seasons, they are less precise than the movements of the stars, but they affect us more directly. Secondly, in space as well as time, variations in earthly phenomena must be noted and followed, as different areas of the earth are suitable in different ways for human use. It is to this point the prince addressed himself, explaining how we must not abuse the special characteristics of four land areas in particular: the rivers, swamps, marshes, and mountains. It should be noted that (according to the prince's speech) Yu did not take a passive role toward nature. Instead he and his assistants raised the mountains, lowered the marshes, dredged the rivers, and banked the swamps, thereby helping nature attain its end.

Finally, earth as well as heaven serves as a guide to correct social behavior. The *Tso chuan* states, "Ritual is the fundamental thread (*ching*) 經 of heaven, it is the true meaning (*i*) 意 of earth, and it is the course of behavior of the people" (Chao 25). This is expanded as follows, "We establish ruler and subject, high and low, and thereby take as pattern the *i* 意 of earth." Tu Yü's commentary explains this as follows: "That ruler and subject have honorable and humble status is patterned after the earth's having high and low." Thus, the mountains and marshes, as well as the stars, provide a pattern for human relationships. That mountains are commonly associated with rulers is suggested by the fact that the word *peng* 崩 means both "death of a ruler" and "collapse of a mountain."

A different analogy from nature is used to the same end in the "Record of Music" in the *Li chi*: "Heaven is honored and earth is humble, and so the relation of ruler and subject is fixed. When the relationship of high and humble has been made clear, then the noble and base receive their status."

We might summarize this as follows: human relationships basically follow a yin-yang pattern, i.e., ruler:subject, father:son, husband:wife. These are a reflection of the fundamental pattern of nature, i.e., heaven: earth, polar star : revolving stars, mountain:marsh. Therefore, even in social relations we use nature as a guide for human conduct.

With this analogy in mind consider the third paradox of Hui Shih, "Heaven is as low as earth; mountains are on the same level as swamps."/35/ On its face it is simply a paradox concerning natural phenomena. As an analogy it is a revolutionary doctrine, suggesting that social distinctions are false. Did Hui Shih have this analogy in mind? It is possible, for his tenth paradox, which is seemingly unrelated to the previous nine, is, "Extensively love the ten thousand creatures. Heaven and earth form one body." Hsün tzu takes pains to refute Hui Shih's third paradox in two different chapters: in "Pu kou" on the grounds that such statements are difficult to grasp and do not concern *li* 禮 and *i* 義 ; and in the "Cheng ming" chapter, where he states that wise men "establish names to indicate actualities—on the higher level to make clear the noble and the base, and on the lower level to distinguish similarities and differences." Thus distinguishing factual differences and differences in social positions are considered together, with the latter receiving primary emphasis by Hsün tzu.

Here then in a number of texts we find a variety of ways in which men are supposed to take nature as a pattern. Together they suggest what the prince may have had in mind when he said that heaven and earth should be used as models for the people. Together they point to an intellectual climate in which an intimate connection between the world of nature and human society was a basic assumption, an assumption which even Hsün tzu could not completely shake.

The cosmology of Prince Chin could be viewed as a philosophy of organism, but it lacks some elements which are prominent in the more fully developed theories of Han times. First, the prince does not mention any portents. According to that theory, human misbehavior will inevitably cause disruptions in the natural order. Portents are infrequently mentioned in the "Chou yü." In one case an earthquake and in another the appearance of a spirit are attributed to improper human conduct. However, portents are not a prominent part of the thought of the "Chou yü," and do not appear in this speech at all. Another difference between Prince Chin's speech and Han cosmology is that there is no mention at all of the five elements of or any of the correlations based on the five elements theory. This is true not only here but throughout the "Chou yü"; while the yin and yang occupy a very important place, there is no mention at all of the five elements.

4. History

The prince appealed not only to the cosmic order, but also to history to support his case. He began his discussion by saying, "I have heard that those who ruled the people in ancient times did not tear down the mountains." He concluded his opening statement by saying, "It was just this to which the sage kings of old were attentive." Then the prince pointed out certain specific "historical" examples. Kung Kung and Kun did not follow the natural order, and they were both destroyed; Yü and the Ministers of the Four Peaks did accord with nature, and they achieved great prosperity. Thus, history is used as a "mirror" for the present, and examples from the past—both good and bad—are used as evidence to support the prince's argument.

The use of history as a mirror was already a venerable practice in China. For example, in the Shao Kao of the *Book of Documents*, the Duke of Chou said, "We shall not fail to mirror ourselves in the lords of Hsia; we likewise should not fail to mirror ourselves in the lords of Yin."/36/ The prince himself quoted from the *Shih-ching*, where King Wen gives the warning, "The mirror of Yin is not distant; it is in the generations of the Hsia lords" (Ode 255.8).

The appeal to history was an almost universal practice among the wirters and speechmakers of late Chou China. Not always, however, could they agree on the historical facts. Concerning the policy of the sage kings toward the natural areas discussed by the prince, there is in the *Kuan-tzu* an account which is quite different from what we have seen in the "Chou yü." According to the *Kuan tzu*, "Shun dried up the swamps and denuded the mountains," and Yü "burned the marshes and swamps" (XXIII, 79). This was justified as follows: it was necessary to "burn the mountains and forests, destroy the marshes, and set fire to the swamps because of the large numbers of wild animals." They had to "denude the mountains and dry up the swamps, because the people were inadequately provided for." In neither the "Chou yü" nor the *Kuan tzu* is any effort made to substantiate the facts as presented. The prevalence of this type of dubious historical argument led Han Fei-tzu near the end of the Chou era to remark that, "it is clear that those who claim to follow the ancient kings and to be able to describe with certainty the ways of Yao and Shun must be either fools or imposters."/37/

The most original aspect of the prince's use of history is his effort to tie the facts together into a systematic pattern. The pattern depicted in the "Chou yü" is one of a gradual rise and a gradual decline in successive dynastic houses. The process is partially described by the prince. In discussing his ancestors of the Chou clan he says, "From the time of Hou Chi, they quieted disorder, but not until the time of Kings Wen, Wu, Ch'eng, and K'ang could they bring peace to the people. From the time when Hou Chi first laid their foundation for making the people tranquil, it was fifteen generations later before King Wen began to bring them order. Only after eighteen generations was King K'ang able to bring them peace. Their difficulties were similar to this. It was fourteen generations ago that King Li first changed the laws." And, "From the time that our ancestors, King Li, Hušan, Yu and P'ing coveted the calamities from heaven, down to today, these calamities have not ceased."

The rise of the Chou clan between the time of Hou Chi and founding of the dynasty is described in greater detail in the first speech in the "Chou yü."

> Formerly our royal ancestors were for generations Lord of Millet,
> serving in that capacity under Yü (Shun) and Hsia. When the

Hsia Dynasty declined, (the Hsia kings) abandoned the Millet (office) and did not concern themselves (with agriculture). Our royal ancestor Pu Ch'u lost his post, and so he hid himself away among the Jung and Ti barbarians. They (he and his successors) dared not neglect cultivated their legacy. They cultivated the teachings and laws. From morning till night they were reverent and diligent in maintaining (their family line) with honesty and sincerity, and serving it with loyalty and faithfulness; for successive generations sustaining their virtue, they did not disgrace their ancestors. Coming to King Wu, he not only manifested the brilliance of his ancestors, but added kindness and harmony to that. He served the spirits and protected the people. There were none who did not happily rejoice in him. The Shang King Ti Hsin was greatly hated by the people. The masses of people could not endure him, and joyfully supported King Wu, resulting in the battle of Mi of Shang. (I, 1)

In the last speech in the "Chou yü" this gradual cycle is applied to successive dynasties. "In ancient times K'ung Chia brought disorder to Hsia, and in four generations it fell. King Hsüan was diligent, and Shang rose to power fourteen generations later. Ti Chia brought disorder to Shang, and in seven generations it fell. Hou Chi was diligent and fifteen generations later Chou rose to power. Yu brought (Chou) to disorder fourteen generations ago" (III, 9).

This consistent pattern enables the prince to generalize on the meaning of history. "Does the King not also have a mirror in the kings of the Li and Miao, and later in the last rules of Hsia and Shang?" "If one investigates the teachings inherited from the former kings and examines their codes of regulations and their criminal laws, and then observes who succeeded and who was destroyed, then all can be known. Those who succeeded certainly had achievements such as Hsia and Lü (that is, Yü and the Ministers of the Four Peaks). Those who were destroyed certainly had failures such as Kung Kung and Kun."

We have seen that the cosmology of the prince is patterned on the yin-yang theory. His interpretation of history has a similar basis, as a cycle of successive rise and fall, similar to the yin-yang alternation of summer:winter, day:night, etc.

From late Chou times there was an interest in systematizing history, but the theories of the different philosophers were by no means in agreement with each other. For example, in the Lü-shih ch'un-ch'iu there is the following theory: "Whenever an emperor or king is about to rise, Heaven will always first manifest some good omen to the common people. In the time of the Yellow Emperor, Heaven made a large number of earthworms and mole crickets appear. The Yellow Emperor said, 'The force of Earth is dominant.' As the force of Earth was dominant, he

chose yellow as his color and Earth as the model for his activities." Then
the Hsia dynasty was correlated with wood, the Shang with metal, and
the Chou with fire. Then "Water will inevitably replace Fire."/38/ After
that water would revert to earth, starting the whole cycle all over again.

Thus, while for the prince the cycle is based on a yin-yang alternation,
in the *Lü-shih ch'un-ch'iu* the succession is based on the succession of the
five elements. For the prince each dynasty represents a complete cycle,
while in the *Lü-shih ch'un-ch'iu* each dynasty, corresponding to a single
element, represents only one phase of a cycle which is not completed until
there have been five dynasties. In the *Lü-shih ch'un-ch'iu*, history and cosmology are more closely integrated than in the "Chou yü," as the dynasty
must correspond with the element which in the natural world is at that
time in the ascendancy. This correlation of the historical and cosmic cycles
makes possible the use of portents, which indicate the rise of a different
element, requiring a change of dynasties. This is something which, as we
have pointed out above, is not found in the "Chou yü." A variation on the
five elements theory was presented by Tung Chung-shu (VII, 23). Specifically, the Hsia dynasty corresponds with red, the Shang with white (metal),
and the Chou with black (water). A different version of the three phase
cycle is presented by Ssu-ma Ch'ien:

> The government of the Hsia dynasty was marked by good faith,
> which in time deteriorated until mean men had turned it into
> rusticity. Therefore the men of Shang who succeeded the Hsia
> reformed this defect through the virtue of piety. But piety degen-
> erated until mean men had made it a superstitious concern for
> the spirits. Therefore the men of Chou who followed corrected
> this fault through refinement and order. But refinement again
> deteriorated until it became in the hands of the mean a mere
> hollow show. Therefore what was needed to reform this hollow
> show was a return to good faith, for the way of the Three Dynas-
> ties of old is like a cycle which, when it ends, must begin over
> again./39/

Ssu-ma Ch'ien's interpretation is like that of the prince in that it is essentially a moral interpretation, where a dynasty's place is determined by a
rise or fall in its virtue. However, it is like the five elements (or three elements) interpretations in that each dynasty represents not a complete
cycle but only a phase: each dynasty here standing not for a specific
element, but for a certain moral quality. There were other arrangements
of the cycle, and the extent of the variations is suggested in the following
passage from Tung, who, rather than insisting on a specific single pattern, argues that a king who founds a dynasty must observe a variety of
cycles, and determine his place in each. "Therefore the king has some
things which he does not change, he has some things which revert after a

cycle of two, some which revert after a cycle of three, some which revert after a cycle of four, and some which revert after a cycle of nine. If one is clear about these things, he will unite heaven and earth, yin and yang, the four seasons, the sun, moon, stars, and zodiacal divisions, the mountains and rivers, and human relationships." (VII, 23). Here the cyclical interpretation of history, which with the prince is a simple yin-yang alternation, receives its fullest interpretation. Note that while there is a difference in the complexity of the theory, the same basic aim remains, that of having human society be in harmony with the cosmic pattern.

The prince's systematizing of history had another function besides correlating the pattern of nature with that of human society. It also served to explain the moral nature of the universe. That the universe is a moral order is an assumption of each speech in the "Chou yü." In one speech this statement is attributed to the former kings: "The Tao of Heaven rewards the good and punishes the evil" (II, 7). All of the speeches in the "Chou yü" have a moral tone; usually they exhort the king to abandon some ill-conceived plan and to return to moral behavior. Usually, also, the king does not heed the good advice of his ministers. And so the speeches end with a note by the narrator that the advice was in fact correct, and that the king suffered as a result of not following the advice. As an example King Hsüan was criticized for not following the custom of the ritual plowing of the sacrificial fields. He refused to reinstate the rites, and in the thirty-ninth year of his reign his army suffered a defeat in the very acres where the ceremony should have been held—thus demonstrating that evil conduct has its consequences.

That virtue is rewarded and vice is punished is a general assumption of ethical philosophers of the period. Sometimes it is heaven or the spirits who favor virtuous action. Hsün tzu sees it more as the working of a law rather than the judgment of a god, but the result is the same. According to Hsün tzu, "Heaven's ways are constant. It does not prevail because of a sage like Yao; it does not cease to prevail bacause of a tyrant like Chieh. Respond to it with good government, and good fortune will result; respond to it with disorder and misfortune will result."/40/

This sounds like an inflexible law, but Hsün tzu was by no means naive enough to believe that one's fortune is invariably a result of one's virtue. Later in the same chapter he says, "The King of Ch'u has a retinue of a thousand chariots, but not because he is wise. The gentleman must eat boiled greens, and drink water, but not because he is stupid. These are accidents of circumstance."/41/ If the universe is a moral order, why are the virtuous often not successful? This dilemma was one which none of the early Confucians were able to resolve.

The philosophy of history in the "Chou yü" provides a way out of this dilemma. We have seen how the rise and fall of dynasties is described in the "Chou yü" not in terms of single good first and bad last rulers, but rather in terms of successive generations accumulating either virtue or evil. Thus, one's virtue or lack of it does have its inevitable consequences, but time is required for the results to appear. This adds a new dimension to the meaning of history. If one looks at a single life, then the good may not prosper, as was the case with the Chou ancestors during the time of the Shang dynasty. However, with the historical perspective, by seeing the world in terms of many generations, the effects of moral causality can be seen.

The moral law also operates in the reverse situation. Those lacking in virtue may for a time enjoy the rewards of their ancestor's virtue, but in time their lack of virtue will have its effects. The prince looks at this in terms again of his own family. "Their difficulties were similar to this. It was fourteen generations ago that King Li first changed the laws. Only after fifteen kings had laid the foundation for virtue did they first bring order. When fifteen kings have laid the foundation for calamity, might it not be beyond saving?" The prince would have become the fifteenth Chou king after the time of King Li, had he lived to succeed to the throne. His interpretation of the moral law becomes poignant because of its personal consequences when applied to his father and himself. He continues, "I am from morning to night alarmed and fearful, and say, 'What virtue may we cultivate, that we may to a small degree brighten the royal house, in order to welcome the blessings from heaven?' "

This interpretation of the historical process could lead to fatalism, but Prince Chin specifically deals with this problem. He pointed out that Yü was the son of the evil Kun, and that the Ministers of the Four Peaks were descendants of Kung Kung, but by changing from the pattern of their ancestors, they were able to prosper. Similarly, Kun and Kung Kung were from noble ancestry, but by turning their backs on their forebears' virtue they brought ruin on themsleves. The prince summarizes: "Of the descendants of those whom heaven has honored, some are working in the fields—this is because they wished to bring disorder among the people. And of the people from the fields, some are at the altars of earth and millet—this is because they wished to make the people tranquil. There is no case which is different from this." Thus, regardless of the actions of one's ancestors, and of the consequent results on one's own life, an individual is not relieved of the moral responsibility for his own actions, nor is one's position ever either completely hopeless or perfectly secure.

In summary, the prince uses the "mirror" of history not only to cite specific cases to support his argument, but also to demonstrate the pattern of history. As this pattern is basically yin-yang, it links the historical process to the cosmic order. It also provides a way of explaining the moral nature of the universe.

5. Rhetoric

Philosophical writing among the Confucians developed from the brief bits of dialogue in the *Analects*, to the more extended debates in *Mencius*, to the topical essays of *Hsün tzu*. In *Hsün tzu* there are many examples of grammatical parallelism and of balancing of phrases. Parallel structure was a feature of Chinese poetry from the beginning; in the late Chou period it was used with increasing sophistication in persuasive prose. In the "Chou yü" parallelism is very prominent, and is used in such a way that it reinforces the cosmological arguments being presented. Let us look at three different sections from the prince's speech to see how this technique is used. In the first the prince has just descrbed Yü's efforts to put each geographic area into proper order, and then he describes the happy results as follows:

t'ien 天 heaven	*wu* 無 without	*fu* 伏 hidden	*yin* 陰 yin
ti 地 earth	*wu* 無 without	*san* 散 scattered	*yang* 陽 yang
shui 水 water	*wu* 無 without	*ch'en* 沈 sunken	*ch'i* 氣 energy
huo 火 fire	*wu* 無 without	*tsai* 災 calamitous	*shan* 燀 blaze
shen 神 spirits	*wu* 無 without	*chien* 閒 divisive	*hsing* 行 conduct
min 民 people	*wu* 無 without	*yin* 淫 licentious	*hsin* 心 hearts
shih 時 seasons	*wu* 無 without	*ni* 逆 violating	*shu* 數 sequence
wu 物 things	*wu* 無 without	*hai* 害 harming	*sheng* 生 life

Here there is a single pattern which is repeated for eight sentences. In each sentence there are four words. The first word is always the subject, the second word is always *wu*, "to be without," and the third and fourth words are the object. The use of a single pattern to apply to the various realms of nature suggests that there may be in fact a single pattern to which they all conform. While grammatically each sentence stands alone, from the context they are seen as pairs. Heaven in the first sentence balances earth in the second, water in the third balances fire in the fourth, spirits in the fifth balances people in the sixth. By the time we reach the last pair, the dichotomy is quite forced, as "seasons" and "things" are not commonly used as a contrasting pair, as the others are.

With the objects (the third and fourth words) there is also pairing, the hidden yin in the first line with the scattered yang in the second, the sunken energy of water in the third line balancing the calamitous blaze of the fourth. Notice that the sunken energy is an example of "hidden yin," while the calamitous blaze is a case of "scattered yang," so that the second pair is a reflection of the first, emphasizing that the pattern of yin-yang duality pervades the universe. Trying to fit the last two pairs of objects into the yin-yang pattern would require some forcing.

One frequently finds in such cases of extended parallelism that the pattern gradually breaks down toward the end. At first I thought that it was simply because the author's ability to invent parallels had passed its limit. Undoubtedly that is part of it, but I think as well that it may be in part a deliberate attempt to go from the obvious parallels to the more subtle, in order to encourage the reader or listener to develop his acuity in discerning the pattern where it is not clearly visible.

In the passage below the prince is telling how the evil rulers of the past failed to discern this pattern:

shang 上 above	*pu* 不 not	*hsiang* 象 imitate	*t'ien* 天 heaven	*erh* 而 and
hsia 下 below	*pu* 不 not	*i* 義 model	*ti* 地 earth	
chung 中 middle	*pu* 不 not	*ho* 和 harmonize	*min* 民 people	*erh* 而 and
fang 方 sides	*pu* 不 not	*shun* 順 accord	*shih* 時 seasons	
pu 不 not	*kung* 供 provide	*shen* 神 spirits	*ch'i* 氣 spirits	*erh* 而 and
mieh 灭 discard	*ch'i* 弃 abandon	*wu* 五 five	*tse* 則 principles	

Here again there is a series of four-word sentences, although the conjunction *erh* is added to make clear that he is talking in terms of pairs. In the first four lines the pattern is very neat. First, a directional word, second the negative *pu*, third, a verb, and fourth, the object. In the fifth line the idea is parallel to the first four lines, but here the grammatical pattern is broken. The sixth line is a summation of the others, thus breaking the pattern completely.

The repetition of the linguistic pattern suggests that no matter where in the universe a person looks, he will see the same pattern in nature.

This is in fact the philosophy of the "Chou yü": there is no instance in the book in which different realms of the universe suggest different courses of action. A person can ignore the order, but if he does look, he will see the same order wherever he looks.

We will look at one more example of parallel structure (see p. 60), from near the beginning of the speech, where the prince was describing the good situation that prevailed when the kings of old did not disrupt the order of nature. Again there is a series of four word sentences, and again they are divided into pairs by use of the conjunction *erh*. The first four lines refer to the geographical areas described by the prince: line one to the mountains, line two to the marshes, three to the rivers, and four to the swamps. Thus the first pair refers to land areas—yang and then yin, and the second area pair to the water areas—also yang and then yin. With the third pair we have shifted from natural areas to human life, with the first line referring to life and the second to death, so that the yang-yin order is preserved. With the fourth pair the four word line is discarded for a different but also parallel structure. Here also the yin-yang relation of the pair is not obvious. With the last line there is no longer a parallel couplet, but rather a single statement. However, still the yang-yin is preserved by the first two words, high and low. Here the implication of the parallel is that if the proper relation between yang and yin is preserved in all the areas listed above, then there will also be the proper relationship between the yang and yin aspects of human society.

The "Chou yü" is by no means the only Chinese text to use parallelism as a means of supporting the argument. Graham stresses this point in his study of "The logic of the Mohist *Hsiao-ch'ü*."/42/ He points out that "At several places in the *Hsiao-ch'ü* one notices the tendency, when faced with two alternative ways of clarifying a problem, to choose parallelism rather than analysis."/43/ He explains this in terms of the nature of the classical Chinese language. "It is natural to connect this choice with the fact that the Mohists were thinking in a language in which the organization of the sentence inflexible resists analysis, yet the parallel organization of paired sentences belongs to the language's ordinary resources. . . . What more natural, then, than that the Mohists should approach logic by seeking tests of the structural similarity of parallel sentences?"/44/ While both resorted to parallelism, the aims of the *Mo-tzu* and the "Chou yü" were different, in that the Mohists were using parallelism as a means of testing the validity of propositions, whereas the author of the "Cho yü" was simply using parallelism as a device for furthering his own argument.

I have talked of the parallelism in the "Chou yü" as if it were a deliberate use of language to support ideas being proposed. Graham has approached it from the opposite point of view, that is that the nature of

chü 聚 accumulation	*pu* 不 not	*ch'ih* 陀 crumble	*peng* 崩 collapse	*erh* 而 and						
wu 物 creatures	*yu* 有 have	*so* 所 place	*kuei* 歸 gather							
ch'i 氣 energy	*pu* 不 not	*ch'en* 沈 sunken	*chih* 滯 congealed	*erh* 而 and						
i 亦 also	*pu* 不 not	*san* 散 scattered	*yüeh* 越 dissipated							
shih 是 thereby	*i* 以 thereby	*min* 民 people								
sheng 生 alive	*yu* 有 have	*ts'ai* 財 wealth	*yung* 用 use	*erh* 而 and						
ssu 死 dead	*yu* 有 have	*so* 所 placed	*tsang* 葬 buried							
jan 然 thus	*tse* 則 then									
wu 無 without	*yao* 夭 early death	*hun* 昬 confusion	*cha* 札 pestilence	*cha'i* 瘥 plague	*chih* 之 's	*yu* 憂 sorrows	*erh* 而 and			
wu 無 without	*chi* 饑 famine	*han* 寒 cold	*fa* 乏 scarcity	*k'uei* 匱 want	*chih* 之 's	*huan* 患 distress	*yi* 以 thereby	*tai* 待 await	*pu* 不 not	*yü* 虞 foreseen
ku 故 therefore										
shang 上 high	*hsia* 下 low	*neng* 能 can	*hsiang* 相 mutually	*ku* 固 firm						

the language determined the argument used, which, extended to the "Chou yü," would indicate that the notion of cosmic pattern is a result of the nature of the language. I would not insist on arguing cause and effect from either direction. There was in the late Chou period a rise of systematic speculation concerning a cyclical pattern which can be applied both to cosmology and history. There was also a rise in parallelism in Chinese expository prose writing. Rather than argue that one development was the cause of the other, I would prefer to take the approach of Tung Chung-shu and say that "Things of the same category activate each other" (XIII, 57)—with the cosmological notions and the rhetorical device strengthening and encouraging each other.

6. Final Note on Dating

In the introduction to this paper I stated my opinion that the speeches of the *Kuo-yü* (including Prince Chin's speech) cannot be reasonably attributed to the time when they were supposed to have been given; it is also unlikely that they were forged by Liu Hsin at the end of the former Han dynasty. Rather, they were probably composed some time during the Warring States period. Let us now review the contents of the speech, to see if they might indicate where we could place the speech in the history of Chinese thought.

The basic concern of the prince is to correlate human society with the order of nature. That there is a close relationship between the human and natural worlds seems to have been an assumption from early times in Chinese culture. For example, many poems in the *Shih-ching* begin with a *hsing* 興 or evocative image. This image is often taken from nature, and sometimes there is a clear parallel between the natural image and the human events that follow, as in Ode 6, between the peach tree and the young bride.

Such analogies are found in the earliest philosophical writing. We have seen the example from the *Analects* in which Confucius compared the relationship of ruler and subjects to that of the North Star and the other stars which revolve around it. In another case (XII, 19) the ruler is compared to the wind and the people to the grass. In *Mencius* there seems to be an almost mystical identity of man and the world of nature, for Mencius says on one occasion that all of the ten thousand things are within him (VIIa, 4), and on another occasion he says that his floodlike *ch'i* fills the space between heaven and earth (IIa, 2).

However, there is not in either the *Analects* or *Mencius* any effort to make a systematic correlation between the human and natural worlds. In fact, the terms yin-yang and five elements do not appear at all in either book.

Apparently, the first philosopher who fully expounded a cosmological theory was Tsou Yen, who, according to Ssu-ma Ch'ien, lived after Mencius. None of Tsou's work survived, but Ssu-ma Ch'ien's biography discusses both his life and his teachings (*Shih Chi*, ch. 74). The earliest surviving work with an orderly treatment of cosmology which can be dated with reasonable certainty is the *Lü-shih ch'un-ch'iu*, which was put together at the very end of the Chou dynasty.

The surviving works are too few in number and too uncertain as to dating to argue conclusively, but the evidence suggests that the problem of cosmology came into focus at the time of Tsou Yen, slightly after Mencius and the *Chuang-tzu*. I would argue from this that the prince's speech belongs to this later period, rather than the sixth century B.C. I think that if this speech had actually been made at that early time, the sophistication of both the language and the ideas would have made the name of Prince Chin one of the most prominent in early Chinese literature and thought.

Could the speech have been written by Liu Hsin? The contents make this very unlikely. We have seen that the cosmology of the prince is based on the yin-yang, but not the five elements. It is not unique in this respect. The "Ta-chuan" of the *I-ching*, which was probably written near the end of the Chou dynasty, is based on the yin-yang rather than the five-elements concept. Furthermore, Hsün-tzu, who lived in the third century B.C., accepted the yin-yang but rejected the five elements theory.

Thus there was a period when the yin-yang and the five-elements theories were not fully integrated. This was changed in the former Han dynasty when the great synthetic works of the era, the Confucian *Ch'un-ch'iu fan-lu* and the Taoist *Huai Nan-tzu* freely made use of both theories. Another idea which separates these works from the prince's speech is the theory of portents, which is not a factor in the speech of Prince Chin.

At the time of Liu Hsin the cosmology of Tung Chung-shu was enjoying its greatest period of influence. I believe that it is very unlikely that Liu Hsin could have written a work in which yin-yang thought figured prominently, but the five-elements concept was ignored. Furthermore, the theory of portents was exceedingly important in the arguments justifying Wang Mang's assumption of power. I find it impossible to believe—if Liu Hsin were indeed a forger trying to help Wang Mang—he would have written (or edited) a speech such as this, a speech which deals with the very problem of dynastic succession but has no reference at all to the succession of the five elements or to portents associated with them. For the speech as it exists is not only not in harmony with the cosmological thought which was prevailing at that time, it also misses a golden opportunity to give justification from an ancient source for the current usurpation.

In conclusion, while we are no closer to knowing who wrote the speech or exactly when he lived, I think a comparison of the ideas of the speech with other early Chinese philosophical writings does provide added evidence to help substantiate the conclusions of Karlgren and Maspero that we are dealing with a work neither of Ch'un-ch'iu nor of mid-Han times, but of the Warring States period.

NOTES

/1/   康有為 , 新學偽經考 , esp. pp. 87–88.

/2/   "On the Authenticity and Nature of the *Tso Chuan*," *Göteborgs Högskolas Årsskrift*, no. 32 (1926): 60–65.

/3/   "La Composition et la date du Tso Tchouan," *Mélanges Chinois et Bouddhiques*, I, 1932. 2

/4/   The Lo was sourth of the royal palace, flowing toward the east. The Ku was north of the palace and it flowed eastward to the Chan, a tributary of the Lo. In this year the Ku overflowed its banks and turned southward, passing along the west side of the royal palace, now flowing into the Lo. The two rivers, flowing together, tore out the southwest part of the wall surrounding the King's palace. The wall was repaired by men from the state of Ch'i in the twelfth month of that year. In the *Ch'un Ch'iu* a flood is mentioned in the seventh month of Hsiang 24. The *Ch'un ch'iu* also mentions a meeting of the Duke of Lu with other dukes in the eight month. The *Tso chuan* says that this meeting was to plan an attack on Ch'i, but that it could not be done because of the floods. The *Ch'un ch'iu* also states that in the winter of that year there was a "great famine," although whether this was on account of the flood is not mentioned.

/5/   It does not become sunk and congealed because there are rivers, and it does not become scattered and dissipated because there are swamps.

/6/   Concerning the legends of Kung Kung, see Bernard Karlgren, "Legends and Cults in Ancient China," *Bulletin of the Museum of Far Eastern Antiquities*, no. 18 (1946): 218–20. Hereafter *BMFEA*.

/7/   See *ibid.*, pp. 250–54.

/8/   Concerning Yü, see *ibid.*, pp. 301–11, and the Ministers of the Four Peaks, *ibid.*, 258–59. I have referred to the Ministers of the Four Peaks as plural, simply because in the prince's speech, they are once referred to as four individuals.

/9/   That is, no creatures such as locusts.

/10/   According to Wei Chao's commentary, the name *Ssu* suggests blessings, and the word *Hsia* means great; hence the two names indicate blessings and abundance.

/11/   Wei explained that *lü* suggests backbone; thus they could act as back-bone for the emperor.

/12/   That is, Yü was descended from Kun, and the Ministers of the Four Peaks from Kung Kung.

/13/   Small states tracing their rulers' lineage to Hsia.

/14/   States with rulers of the Chaing clan.

/15/   Kun was descended from the Yellow Emperor, and Kung Kung from Yen Ti (Shen Nung).

/16/   That is, Yü and the Ministers of the Four Peaks.

/17/   Ode 257.2.

/18/   Ode 257.11.

/19/   The fifteenth king after King Li would have been the prince himself, had he lived to succeed to the throne.

/20/   Observing heaven, earth, people, seasons, and spirits.

/21/   Ode 255.8.

/22/   Wei takes "patterns" to refer to heaven and markings, *wen* 文 , to refer to the written classics. But *wen* could also refer to the markings on birds and breasts, as in the "Ta chuan," II, 2.

/23/   King Ling's successor, r. 544–519 B.C.

/24/   Chen Ting, the fourth king after King Ling; r. 468–440.

/25/   *Sourcebook in Chinese Philosophy*, Princeton, 1963, p. 784.

/26/   James Legge, *The Chinese Classics*, vol. 3 (Hong King, 1960) pp. 312–13 (my translation).

/27/   XX, 8. I take the original 象天 as 像 and 天 , following Wei.

/28/   Joseph Needham, *Science and Civilization in China* (Cambridge, 1970), 188.

/29/   Legge, *Chinese Classics*, 3: 18.

/30/   Bernard Karlgren, "*The Book of Documents*," *BMFEA*, no. 22 (1950): 3.

/31/   *Science and Civilisation*, 2: 557.

/32/   *Ibid.*, p. 562.

/33/   Karlgren disagrees with this interpretation. See "Glosses on the Book of Odes" no. 768 (Stockholm, 1964).

/34/   *Chou-i* VIII, 32a. Chang Heng cited from Needham, *Science and Civilisation*, 3: 217. For discussions of early Chinese astronomical instruments see Needham, 3: 284–390, and Henry Maspero, "L'Astronomie chinoise avant les Han," *T'oung Pao* 26 (1929):267–356.

/35/   In *Chuang tzu*, X, ch. 33.

/36/   Karlgren, *Book of Documents*, p. 49.

/37/   Sec. 50. Burton Watson, trans., *Han Fei Tzu: Basic Writings* (New York 1964), p. 119.

/38/   XIII, 2. tr. Wing-tsit Chan, *Sourcebook in Chinese Philosophy*, p. 250.

/39/   Burton Watson, *Records of the Grand Historian of China* (New York, 1961), 1: 118.

/40/   Sec. 17. Burton Watson, trans., *Hsün Tzu: Basic Writings* (New York, 1963), p. 79.

/41/   *Ibid.*, p. 83.

/42/   *T'oung Pao*, ser. 2, vol. 51.

/43/   *Ibid.*, p. 50.

/44/   *Ibid.*, p. 49.

The Concept of Change
in the *Great Treatise*

Gerald Swanson

Introduction

The *Great Treatise*, or *Ta-chüan*, is a separate book which from Han times has been appended to editions of the *I Ching* or *Book of Changes*./1/ The *Great Treatise* is the largest and probably the most important of seven texts known in the *I Ching* commentatory tradition as the "ten-wings."/2/ The ten-wings are composed of true commentaries, such as the *t'uan, hsiao-hsiang,* and *wen-yen*; glosses on the names of the hexagrams or their symbolic correlations such as the *ta-hsiang chüan, shuo-kua, hsü-kua* and *tsa-kua chüan*; and diverse collections of essays about the nature of the *I Ching*, such as the *Great Treatise* and the first two chapters of the *shuo-kua chüan*. Many of these are undoubtedly older than the *Great Treatise*, but the exact date of any of them is very difficult, if not impossible, to ascertain.

The *Great Treatise*, then, is a diverse collection of essays written about the *I Ching*. As such, it is the beginning of scholarship about the *I Ching* in China. It is divided in the traditional style into two parts. There are twenty-four chapters, twelve in each part. There are 887 lines in the traditional "breath stop" punctuation by Han K'ang-p'o and 4,465 characters in the entire text./3/ This places the *Great Treatise* in about the same relative size as the *Tao-te ching*.

Questions of authorship and dating seem at this time to be insoluble. The traditional ascription to Confucius is no longer held outside of China, and many critical Western scholars feel the text comes from an unknown author of the early Han./4/ The first piece of hard evidence we have in dating the *Great Treatise* is a quotation of it in the *shih chi*. There it is quoted in an essay presumably written by Ssu-ma T'an, so we know it cannot be later than 90 B.C./5/

In terms of content, the *Great Treatise* has most in common with the *t'ien ti, t'ien tao,* and *t'ien yün* chapters of Chuang tzu. These chapters, although certainly not written by Chuang tzu himself, are most probably late Chou texts.

In terms of grammar, the *Great Treatise* clearly conforms to what is thought to be a late Chou "literary language."/6/ It manifests a grammar similar to the *Lü-shih ch'un-chiu, Chan-kuo ts'e, Hsün tzu* and *Han-Fei tzu*, but is closer to the later chapters of *Chuang tzu*. Although the *Great Treatise* has more in common with certain late chapters of the *Chuang tzu*, that does not make it a Taoist text. Rather, I suggest a separate group of scholars with their own specific interests put together this rather diverse collection known collectively as the *Great Treatise*.

The Problem of the *Great Treatise*

At about the time the *Great Treatise* was written, a number of Chinese thinkers became interested in the early origins of their culture. There abounded a number of mythic sovereigns for which sacrifices were still being conducted in the third or second century B.C. These sovereigns were most probably local deities, but as they found their way into pre-*Great Treatise* texts such as *Chuang tzu* and certain chapters of the *Kuan tzu, Hsün tzu, Tso-chüan*, and *Han-fei tzu*, they achieved a certain legitimacy that derived from a reverence for any ancient text./7/

Part of the problem for the authors of the *Great Treatise* was to systematize this number of mythical kings into a generally acceptable orthodox line of transmission as well as provide all the details necessary for a cultural history. In the process of this systematization the authors of the *Great Treatise* changed whatever was known of these early deities into a coherent cultural history. In the *Great Treatise* these early kings were not merely kings; they were divine. And being divine, they provided the Chinese people with the tools for a culture that separated them from non-Chinese peoples. Their divinity rested upon their being somehow intermediate between the bipolarities of heaven and earth. Thus the wisdom of Fu Hsi and all the others rested partially upon the notion of perspective. They were simply in a position to better observe the patterns of heaven and the patterns of earth and from that created the numerous accoutrements of a physical culture.

Another problem for the authors of the *Great Treatise* was to preserve and synthesize two early schools of *I Ching* interpretation: the hexagram and trigram schools. Both of these interpretative methods go back to the very earliest texts we have in the Chinese literary corpus./8/ And with the possible exception of the *t'uan chüan*, or "Commentary on the Decision," they are held apart in the various traditions of the other ten-wings. We find in the *Great Treatise* essays the hexagram and trigram methods of interpretation used inextricably together, especially in the cultural history chapters.

But we, as students of Western civilization, are not always interested in struggles over the orthodox transmission of mythic sage kings or in the systematization of pre-Han cults associated with these deities. We are interested in the broad set of presuppositions of which the text makes use and assumes rather than defends, especially those which are somehow similar to presuppositions in Western philosophy.

The methodology for this essay was taken from Alfred North Whitehead:

> When you are criticizing the philosophy of an epoch, do not chiefly direct your attention to those intellectual positions which its exponents feel it necessary explicitly to defend. There will be some fundamental assumptions which adherents of all the variant systems within the epoch unconsciously presuppose. Such assumptions appear so obvious that people do not know what they are assuming because no other way of putting things has ever occurred to them. With these assumptions a certain limited number of types of philosophic systems are possible, and this group of systems constitutes the philosophy of the epoch./9/

The Technical Terminology of Change

Every student of spoken Chinese learns that the common word for "change" is *pien-hua*. This term is a combination of two nouns, called a binome, both of which presumably mean change. In the modern language there seem to be no separate meanings attached to either of the binomial terms in the expression. But can we be sure that this was the case in the late years of the Warring States period?/10/ The term *pien-hua* is a fairly late phenomenon in Chou texts. It does not occur in the *Analects* nor the *Meng Tzu*, but is found regularly in both the *Hsün Tzu* and the *Chuang Tzu*.

There is, however, no extended argument concerning the notion of *pien-hua* in the *Great Treatise*. How, then, are we to reconstruct the concept of change or its general set of presuppositions? Although there exists in the *Great Treatise* no extended discussion of the view of change which we hope to reconstruct, there are very definitely technical statements made about change and its variations. These statements are found *en passant* to other arguments in the text. This leads to the suggestion that perhaps the best way to understand the metaphysical presuppositions of this text, is *not* to study the points about which the authors of the *Great Treatise* differ with other texts of the period, but to examine that which they assume as given. The notion of change in the *Great Treatise* is an example of an assumed presupposition. The concept of change in this text is *not* a philosophical issue; it is not argued for, but rather it is assumed and used in support of other arguments. What, then, was the philosophical problem or problems which faced the authors of the *Great Treatise*? Part of their concern, if not the whole of their concern, was to give a philosophical account of the legacy of the *Book of Changes* as they had received it, in terms of the dominant philosophy of their time. For metaphysics and natural philosophy this meant the *yin-yang* speculation deriving from Tsou Yen and his school. And since little of Tsou Yen's corpus has come down to us, the *Great Treatise* represents a major source of this type of speculation. The authors of the *Great Treatise* did not merely interpret the *Book of Changes* in terms of *yin-yang* speculation. Rather, they took it as a presupposition and constructed explanations which were applications or modifications of these presuppositions. But the legacy of the *Book of Changes* that the authors of the *Great Treatise* had received had employed bipolar opposition and implicit analogy long before there were technical terms like *yin* and *yang* to describe it. This leads us, then, to ask what is *distinctive* about the *yin-yang* application at the hands of the authors of the *Great Treatise*; that is, how did it differ from the prephilosophical use of bipolar opposition and analogy that went back as far as the Shang dynasty oracle bones?

The hexagrams of the *Book of Changes* contain bipolar oppositions of straight and broken lines; the two trigrams can also be thought of as a bipolar opposition. If the authors of the *Great Treatise* are giving philosophical expression to prephilosophical attitudes in the structure of the hexagram, then we should expect their technical terminology to some degree at least follow the operations that occur in the hexagram. We will thus want to correlate, where possible, technical terminology in the *Great Treatise* concerning the seemingly accepted concept of change with aspects of the hexagrams which derive from a far earlier period.

With these necessary preliminaries set aside, then, we shall attempt to first isolate the technical terms for change as they occur in the *Great Treatise*. We shall do this first by observing what appear to be single-term definitions of these technical terms and then see how they are used in conjunction with other technical terms. We shall also examine the extent to which this terminology correlates to aspects of the hexagrams.

We shall begin by examining the single term uses of *pien*. Perhaps the most illuminating single term use of *pien* occurs in chapter A, 11: "Therefore closing the door is called *k'un*; opening the door is called *ch'ien*. Now closing, now opening is called *pien*."/11/ There are a number of comments which may be made concerning the above statement. First, it appears to be a technical definition. *Pien* is being defined by virtue of the fact that something x "is called" *pien*. Secondly, the context of the definition definitely excludes other terms such as *hua*: this is an attempt at a precise definition of *one* term. If that is the case, then, what can be made of the definition? In the first place, *pien* is here seen to be a combination of one instance of *ch'ien* and one instance of *k'un*. Now if we seek to find a correlation with this definition and aspects of the system of the *Book of Changes*, there are two possibilities. First, *ch'ien* and *k'un* are the names of the first two hexa-grams. The other possibility is the application of *ch'ien* and *k'un* to the lines of the hexagram. The metaphor of the gate is used again in chapter B, six: "The Master said: 'Are not *ch'ien* and *k'un* gateways to the changes? *Ch'ien* is a *yang* thing; *k'un* is a *yin* thing. When the *yin* and *yang* unite their powers the hard and the soft obtain form."/12/ The reference of *ch'ien* to *yang* and *k'un* to *yin* presents the second possibility of interpreta-tion. In this possibility, the definition of the term *pien* is associated with the fact that the lines of the hexagram have two and only two possibilities— either a broken or an unbroken line. Each line must have one of the two and there is no third possibility. Therefore translate the term *pien*, in its technical sense, as alternation./13/

At this point it is important to notice that *pien* is a fundamental alter-nation of *yin* with *yang* and *yang* with *yin* with no other possibility. This is not the same as the term "alteration." Alteration implies a change of quali-ties, as when a tailor alters a suit. With alteration there is the implication of something that persists through the change, in this case, the suit. But the qualities of which there is an alteration may be many. The suit may be too big, too small, or have any number of qualities, both of which mutually exclude one another. Moreover with *pien*, conceived as the alternation of opposites, there is no emphasis upon that which subsists through the change. The six lines are always filled with a *yin* or *yang* line and the notion of an empty slot with no content does not exist in the *Book of Changes.*

We turn now to other examples of single term uses of *pien*, conceived as alternation of opposites. In chapter B, 1, we have: "When the hard and the soft displace each other, *pien* is located within their midsts."/14/ Again we have alternation of hard and soft, aspects of the *yang* and *yin* lines. Another example is from chapter B, 8: "The hard and the soft exchange and cannot form a rule or regulation. They are only in accord with *pien*."/15/ Again alternation of bipolar opposites seems to be the proper interpretation of this term.

We turn now to the term *hua*, translated as transformation./16/ The only definition of *hua* that we have in our sources is found actually in a single term definition of alternation. The location is chapter A, 12: "There-fore what is beyond form is called *tao*. What is subsumed under form is called vessels. Transforming and measuring is called alternation."/17/ A restatement of the same notion occurs later in the chapter, "To transform and measure rests with alternation." Thus transformation, in the *Great Treatise* at least, is seen by its authors as a species of change such that when measured, that is, reduced to order, becomes alternation./18/ But clearly not all changing things in the natural world can be reduced to the interac-tion of opposites. The concept of transformation, then, is used to explain the transformations of things which change form, such as a butterfly changing from a silkworm./19/ But if *hua* when reduced to order becomes alternation, who or what is it that carries out the reduction? The answer to this question is found in chapter A, 8: "The sages had the means to perceive the mysteries of the world and so measured their form and appearance. They made representations of their material fitness and for this reason called them images. The sages had the means to perceive the movements of the world and so observed their convergence and development. They thereby put into practice canons and rituals. They appended texts in order to determine good or evil fortune."/20/

In this quotation the sages are not measuring or reducing to order the type of change in the natural world which is basically a change of substance. The sages, are, however, measuring the "mysteries of the world." This leads to a further complication of our understanding of the concept of transformation: the association of transformation with the term *shen*, the supernatural, divine, and spiritual.

In the famous cultural history chapter, chapter B, 2, we have a continuation of the discussion of the contributions to Chinese society by the sages: "When Shen Nung died, Huang Ti, Yao, and Shun arose. They understood the alternations and caused the people not to grow weary. Being divine they transformed them and caused the people to find what was suitable."/21/ The point of interest here is that the sages were divine and that there was *some* association between this divinity and the concept of transformation. Another association between the sages and the term transformation occurs in chapter A, 4: "Therefore the sages did not transgress. They were active in all directions and yet did not go to extremes. They enjoyed heaven; they understood destiny, therefore they were able to love. They encompassed the transformations of heaven and earth and did not go to excess."/22/ Another important instance of the association of *shen* with *hua* occurs in chapter B, 5: "To know all about the spiritual and understand transformation is the fullness of power."/23/ These quotations are enough to show, I suggest, that there was some correlation between *hua* and *shen*, or between transformation and the spiritual, divine, supernatural, or unknowable.

This puts us in a position to consider the relation between *hua* and its association with *shen* and alternation. In chapter A, 5, we have: "What *yin* and *yang* do not fathom, is called the supernatural."/24/ Transformation, then, insofar as it is correlated in the minds of the authors of the *Great Treatise* with the supernatural or the spiritual, stands opposed to alternation. The alternation of *yin* and *yang* opposites constitutes one type of change; the change of substance which does not fit the mold of bipolar opposition constitutes another type of change. If alternation is "ordered change," then it is probably correct to say that transformation is that term which stands for all the unordered, random, or chaotic change we find in the natural world.

We may now conclude our discussion of the concept of transformation by returning to our original problem: the significance in the *Great Treatise* for the term *pien-hua*, or "alternation and transformation." The problem citations are from chapter A,1, and chapter A,2: "When images are created in heaven and forms are created on earth, alternation and transformation appear." "The hard and the soft displace each other and produce alternation and transformation."/25/ In both cases alternation and transformation are bipolar opposites. Alternation is ordered change of bipolar opposites; transformation is that which is not reducible to bipolar opposition. Transformation has the derivative notions of change of substance, random, unexplainable change, or chance./26/ And finally we must observe that for the principle of order there is a spiritual correlate which is not reducible to order.

There are two other important technical terms for change in *Great Treatise*: the notion of change as it occurs in the title of the book, the *I Ching*, or *Book of Changes*, and the notion of *t'ung*, translated here as a development, is intimately associated with the term alternation and is a term which finds correlation in the structure of the hexagam./27/ Two single term definitions of *t'ung*, or development, occur in the *Great Treatise*, one in context with the single term definition of alternation. In chapter A, 12, we have: "Therefore what is beyond form is called the tao. What is subsumed under form is called vessels. Transforming and measuring is called alternation. Displacing and setting into motion is called development."/28/ We have in this citation two consecutive technical terms for change. As we have seen above, alternation is defined as the interaction of bipolar opposites. Development as it relates to the hexagram is basically *an expansion* of these bipolar opposites to fill the six line matrix. Any line in the hexagram may be either broken or unbroken, but the hexagram is composed of six lines, not just one. The expansion of these bipolar opposite possibilities to fill the pattern of lines associated with any specific hexagram was, of course, not random. It was important enough in the minds of the authors of the *Great Treatise* to warrant a specific technical term to describe this expansion.

Another single term definition of development occurs in chapter A, 11, again in a context with the term alternation: "Therefore closing the door is called *k'un*; opening the door is called *ch'ien*. Now closing, now opening is called alternation. Going and coming without limit is called development."/29/ Again development is intimately associated with the term alternation. Here, as before, development is an expansion of bipolar opposition. The terms "going and coming" are bipolar opposites; the displacement of one by the other is called alternation. Two separate instances of alternation in the same hexagram represent the beginnings of development of a pattern of lines. When this is extended to fill the six-line matrix we have a hexagram. Alternation of broken and unbroken lines without limit eventually yields the 384 lines of the complete system known as the Changes. But our purpose here is not to explain the inter-action of the technical terms, but only to isolate them. We have yet to deal with the most general term in the *Great Treatise* for change—the term for which the text is named.

The final term we need to consider in the technical terminology of the concept of change is the term *i* translated here as "change." This is the most common term for change in general as used by the authors of the *Great Treatise* and is the term employed in the name of the classic known as the *Book of Changes*. There has been much speculation on the etymological origins of the term *i* by previous scholars who have con-cerned themselves with the *Book of Changes*./30/ The isolation of the meaning from the contexts in which it is used has not, however, been explored as a method of gaining insight into the early meanings of the term.

Probably the first occurrence of the term in what we would call an "interesting" context occurs in the *"Hung-fan"* chapter of the *Book of Documents*; the context is the conversation of Chi tzu, a disaffected Shang prince, with King Wen: "He said: What the king scrutinizes is the year, the dignitaries and noblemen the months, the many lower officials the days. When in years, months and days the seasonableness has no changes, the many cereals ripen, the administration is enlightened, talented men of the people are distinguished, the house is peaceful and at ease. When in days, months and years the seasonableness has changed, the many cereals do not ripen, the adminstration is dark and unenlight-ened, talented men of the people are in petty positions, the house is not at peace."/31/

When we examine the occurrences of the term *i*, or change, in the *Great Treatise* several observations can be made; first, the term seems to be associated with *ch'ien* and *k'un* and the relative interaction of heaven and earth. Secondly, the term is associated with life and living things. We will need to discuss several of the paradigm uses of the term with each of these associations before we can make inferences concerning the relation of the term to other technical nomenclature.

To examine the first set of associations, *ch'ien* and *k'un* and heaven and earth, we can look at two different citations at the same time: the first is from chapter A, 7, and the second is from chapter A, 12: "Heaven and earth establish positions and change moves in their midsts"; "Are not *ch'ien* and *k'un* the secret of the Changes? When *ch'ien* and *k'un* form ranks, changes are set up within their midsts. If *ch'ien* and *k'un* are destroyed then there is no way to observe change. If change cannot be seen, then *ch'ien* and *k'un* may almost stop."/32/

The interesting point in these quotations from separate locations in the *Great Treatise* is the association of change with the forming of ranks or the establishment of positions. The first quotation argues from the point of view of the macrocosm. Change in the natural world moves through the matrices of heaven and earth. This is not to say that *all* change is alternation of bipolar opposites, somehow imitative of a set of paradigms called heaven and earth. For all change is not systematic change.

In the second quotation, the word change occurs twice. The first instance is a reference to either the text known today as the *Book of Changes* or perhaps to an oral tradition known as "the Changes." The second instance of the term is seen from the perspective of the interaction of *ch'ien* and *k'un* or the creative and receptive principles of nature. But here the authors of the *Great Treatise* are clearly not referring to the creative and receptive principles as they might occur in the macrocosm to an observer, but to *ch'ien* and *k'un* as they occur in the system of thought known as the Changes. Thus we have a possible equivocation between the separate uses of the term change. Change, when used from the point of view of the macrocosm, or change in the natural world, is not the same thing as change in the microcosm, or change as it is employed within the context of the system of thought known as the *Changes*. That is, change is used here in two separate senses: change as a nontechnical term in the natural world and change considered as a technical term within the *Book of Changes*. As a nontechnical term change is a combination of systematic change and nonsystematic change, that is, in a nontechnical sense, change *is* alternation and transformation. In the other sense of the term, change as a technical term within the system of the *Book of Changes*, change is a combination of alternation and development. But the concept of change, even in its nontechnical sense, is a combination of bipolar opposites.

A few words are perhaps in order to conclude this section of the technical nomenclature of the concept of change in the *Great Treatise*. First, the evidence in our sources is hardly such that we can conclusively settle all the questions which face us. But it is clear that the authors of the *Great Treatise* are using some terms in a very specific and technical sense. In general we should expect some correlation between those technical terms and the various operations or aspects of the hexagrams. Alternation and development are the only two terms which have structural correlations in this sense and thus are the two most important technical terms in the concept of change as understood by the authors of our text. Alternation is the alternation of bipolar opposites; this occurs in the *yin* and *yang* line opposition of the hexagram. Development is the development, extension, or penetration of these bipolar oppositions into the six-line matrix of the hexagram. The term *i*, as it occurs in the title of the text, the *I Ching*, is a combination of alternation and development. The term *i*, as we have seen from the "Hung-fan" citation, also has some connection with time and the duration of regularities taken from the observation of nature. The term *i* in a nontechnical sense is a combination of systematic and nonsystematic change, or a combination of *pien* and *hua*. *Hua* or transformation is a term for the preanalytic and pre-reductive aspects of change which contain spiritual and supernatural correlations. Transformation when reduced to order becomes alternation. With the isolation of a set of technical terms, we are now in a position to examine the concept of change in the *Great Treatise*.

Developing Alternations in a Time-Relation Continuum

If by the concept of change we mean the way change is explained, any essay dealing with the explanation of change in the natural world must eventually deal with the concepts of time and space and problems concerning these ultimate categories of experience. All change must occur in time and space and the dimensions of these concepts will intimately affect any understanding of change./33/

We have seen in the previous section that the two technical terms for change for which there is something of a structural counterpart in the hexagram are "alternation" and "development." Since a basic presupposition of the *Book of Changes* is an analogy between the macrocosm of the world and the microcosm of the system of relations of the hexagrams, we will want to examine both aspects of the analogy. We shall also want to examine how concepts of space and time are woven into the system of the *Book of Changes* and the effect of both these concepts upon the notion of change.

The correlation of the technical terms alternation and development with time, in the form of the four seasons, is made twice in the *Great Treatise*; the first occurrence is chapter A, 6: "Alternation and development correspond to the four seasons."/34/ And again in chapter A, 11: "As to alternation and development, nothing is greater than the four seasons."/35/

These quotations seem to be saying the same thing from different perspectives. The first emphasized the macrocosm, in the sense that alternation and development "correspond" to the four seasons. The relation of microcosm to macrocosm is stated several ways in the *Great Treatise* and the notion of correspondence is used only rarely. The most common relations between the two poles of the analogy are as some form of imitation. In that sense our citations have the sense of "alternation and development imitate the four seasons." In the second instance the emphasis seems to be upon the terms alternation and development. The second quotation states that the seasons of the year are the best examples of alternation and development. At this point we must make some attempt at interpretation to get at the meaning of these statements.

Clearly the authors of the *Great Treatise* are not implying that the technical terms *are* the four seasons, that is, the technical terms as such are not imitations or correspondences of the four seasons per se; rather the relation of the technical terms to each other imitates or corresponds to, not the four seasons themselves, but aspects of their relations to each other or aspects internal to themselves. We must infer, as best we can, what the authors of the *Great Treatise* had in mind when they posited the relation between the terms of both the microcosmic relata and the macrocosmic relata. We are on reasonably firm ground with respect to the microcosm: alternation occurs between broken and straight lines as a basic relation between creative and receptive processes. This alternation is *extended* from the possibility of one or two bipolar opposites to the possibility of six of twelve bipolar opposites. In other words, the hexagram is *not* composed of six unrelated slots containing either a straight or broken line. The hexagram is a developmental pattern of symbols, the content of which is, in each instance, a bipolar opposition.

If we consider the macrocosmic relata of the four seasons, we may infer two things. First, the four seasons, or any unit of time in general, contain alternations of bipolar opposites. Secondly, together with the set of relations internal to them, the development of these alternations forms a unique pattern. Thus change in the natural world occurs between opposites such as wet and dry and cold and hot, but movement of opposites is not random. Winter, we can infer, is not characterized by the haphazard development of the opposites of hot and cold. Rather there is an *orderly pattern* of development.

We find a similar orderly pattern of development in the so-called twenty-four solar terms. In the following table each of the four seasons is divided into six divisions, producing a very interesting comparison with the six lines of the hexagram. We notice that summer and winter contain the opposites of lesser and greater heat and cold, while spring and autumn contain opposites of rain and dew respectively./36/

Thus far there are three categories of explanation: alternation, development, and time. The role of time in this scheme is very important. First of all, time is circular and not linear. Time in general is a hypostatization of the four seasons. These four units of duration had their own developmental patterns based upon the interaction of opposites. But time became a principle of organization of these alternations and their extensions such that there was now a *boundary* or a *limit* to this development. The alternations could no longer proceed into infinity. The principle of cyclicity generalizes particulars in the sense that the recurrent cycles bring to our attention aspects of regularity that might otherwise be lost or gone unnoticed. Thus by dividing duration into the four sections, or perhaps, stated more correctly, hypostatizing a general notion of time from the recurrent cycles of the four seasons, we are able to observe the

interaction of opposites more clearly. There is a certain simplicity and elegance in the relating six variables which might be lost or gone unobserved if the general notion of time had been taken from a metaphor of the year.

Secondly, time is a principle of organization in the sense that it plays a role of a receptacle. In the five elements theory, each of the five elements was thought to dominate over the others for roughly the same duration. Likewise each of the four seasons, with their specific patterns of alternations, dominated over the others and lasted only so long. Perhaps the best method for explaining this simple yet difficult to describe conception rests in the use of metaphor. Time is a receptacle, a container, or a room in which events happen and relations occur. As a principle of organization it gives coherence to all that occurs within it./37/

We return briefly now to the microcosmic relata. To what extent are the notions of alternation and development in the hexagram related to the concept of time? On this subject the authors of the *Great Treatise* are very explicit. In chapter B, 1, we have: "Alternation and development are things which tend toward the proper time."/38/ This statement is incomprehensible without reference to the macrocosm-microcosm analogy. The four seasons are the macrocosmic relata, but to what in the microcosm are the analogues? The answer is obviously the six-lined matrix of the hexagrams. The hexagram, then, in some yet unspecified manner is a system analogue for a general conception of time. We have seen that time, conceived in a nonlinear fashion as the hypostatization of the progression of the four seasons, provided the notion of a receptacle as well as a boundary for the developing alternations. Also from the notion of cyclicity as such we observe aspects of completeness, richness of relations, and equality of relata./39/ These aspects we need not project upon the hexagram; the authors of the *Great Treatise* and other commentaries of the ten-wings are quite clear about the matter. In the first place, the notion that the hexagram and the emerging or holistic conception which comes from the verbal and nonverbal images of its lines constitutes "time" of a sort is much older than the *Great Treatise*./40/ In the *t'uan* or "Commentary on the Decision" for the first hexagram we have: "Because the holy man is clear as to the end and the beginning, as to the way in which each of the six stages completes itself in its own time, he mounts on them toward heaven as though on six dragons."/41/ Also the *Wen-yen* Commentary for the same hexagram is very revealing although the word "time" is not used: "The six individual lines open up and unfold the thought, so that the character of the whole is explained through its different sides."/42/

Thus the concept or image which emerges from the whole and which is therefore identified with the name of the hexagram is called "time" of the hexagram. Looked at in this way, the sixty-four hexagrams can be conceived as sixty-four divisions of time./43/ To be sure, the authors of the *Great Treatise* are not restricting themselves to the framing of a physical theory such as the five elements. But the *yin-yang* theory of Tsou Yen could account for both physical and psychological phenomena. And yet the thinkers who develop the *Book of Changes* and the authors of the *Great Treatise* who tried to weave an earlier legacy into a coherent system went beyond *yin-yang* speculation. The creative and receptive are concerned with life and life as an ongoing organic process. The emergent conception of time, then, reflects these preoccupations. Further, this notion of time is not an a priori form of intuition as it was for Kant. It is, rather, a category of apperception, that is, it is a concept used to explain and emphasize certain features of the perception of duration and deemphasize others. As a hypostatization of the four seasons, time plays the role of a container or receptacle of all change. But the notion of time is supplemented with another category of apperception.

This second category is not space, as we would imagine, but *wei*, rank, or relative position./44/ The notion of relative position originates from the apperceptive division of heaven and earth and their correlation to high and low. The general concept of *wei*, rank or position relative to a whole (or some context), is an extension of this bifurcation. The importance of relative position is brought out quite early in the *Great Treatise*; in chapter A, 1, we have: "Heaven is exalted; earth is lowly. Thus *ch'ien* and *k'un* are determined. Because they are arranged according to the low and the high, the honorable and the mean take their relative positions."/45/ Here we see the notion of value derived from rank or relative position. *Ch'ien* and *k'un* are images of heaven and earth; the correlations of high and low are associated with *ch'ien* and *k'un* by their analogy to heaven and earth. Value judgments associated with *ch'ien* and *k'un* are *not* based upon static intrinsic value. That is, *ch'ien* is not in every case honorable while *k'un* is lowly. As the text says: "The honorable and the mean take their relative positions," in accord with the low and the high. *Ch'ien* is not always high, nor is *k'un* always low. In fact, in a cosmological system that claims completeness and equality of relata, *ch'ien* and *k'un* will share all values equally./46/

We shall now ask the question, to what extent did the early Chinese experience the world in terms of rank or a general notion of relative position? That is, to what extent is the macrocosm infused with apperceptions of relative position? Perhaps the most interesting source for some insight into the question is found in the *t'ien-tao* chapter of Chuang Tzu:

> The ruler precedes, the minister follows; the father precedes, the
> son follows; the older brother precedes, the younger brother follows; the man precedes, the woman follows; the husband precedes, the wife follows. Honor and lowliness, precedence and

following are part of the workings of Heaven and earth, and from them the sage draws his model.

Heaven is honorable, earth lowly—such are their ranks in spiritual enlightenment. Spring and summer precede, autumn and winter follow—such is the sequence of the four seasons. The ten thousand things change and grow, their roots and buds, each with its distinctive form, flourishing and decaying by degree, a constant flow of change and transformation. If Heaven and earth, the loftiest in spirituality, have yet their sequence of honorable and lowly, of preceder and follower, how much more must be the way of man./47/

Here the author of this *Chuang tzu* passage is clearly arguing by analogy from the sequences in nature to the sequences or relative positions in human institutions./48/ But *Chuang tzu's* analogy and the previously quoted analogy from the first paragraph of the *Great Treatise* are based upon the relation of opposites to each other, especially with reference to sequence and value. There is no reference here to any sixfold division of relative position. The sixfold division of relative position is *derivative* from and based upon the primary relation of two variables, heaven and earth./49/ For a sixfold division we may return to the twenty-four solar terms. In this schema we may now see the two categories of time and relative position working together. The six divisions of each season are based upon expansions or development of the type of bipolar opposition found in *Chuang tzu* and the opening statement of the *Great Treatise*. The totality or unity of these six divisions is represented by the totally different conception of time which emerges from them. In other words, the concept of the seasons is not the *sum* of the six divisions; nor is the concept of time the *sum* of the six relations. The concept of time is different *in kind* from the concept of relation or relative position. The two concepts are derived from each other in that time "emerges" from the development of alternations or that bipolar alternation and their extension or development "participates" in the time. And whether the metaphors of emergence and participation are changed for something else, time and relative position remain intimately interconnected with each other. For instance, spring and winter both share a similar form in terms of development of bipolar opposition, and yet, since the content of the oppositions are different, the emergent conception of the time is different.

We turn now to the microcosmic analogue of relative position in the hexagram. The *Great Treatise* speaks of the notion of relative position of certain lines. In chapter B, 9, we have: "The second line and the fourth line have the same function, but have different positions. Their goodness is different. That the second line is mostly praised and the fourth line is mostly feared is because the fourth line stands near the ruler of the hexagram . . . The third line and the fifth line have the same functions, but different positions. That the third is mostly unlucky and the fifth is mostly meritorious is due to gradations of honorable and mean positions."/50/

Here the discussion is concerned with the relative positions of lines three and five and lines two and four. The point of interest here is that only certain line relations were considered important. There are two types of relative position postulated by the commentary tradition in the ten-wings. First, the relation of any line to its position in the six-lined matrix of the hexagram. This type of relation we might call "matrix relative." There are two subvarieties of matrix relative relations. First, lines two and five, which are the central lines of the lower and upper hexagrams respectively, are always central irrespective of whether they are *yang* or *yin* lines. This is normally, but not always, a good relationship. The second type of "matrix relative" relation is called "correctness." Any given line is "correct" or "incorrect" by reference to the specific line configurations of hexagram sixty-three. Thus the model for correctness requires *yang* lines in the first, third, and fifth places and *yin* lines in the second, fourth, and sixth places.

The second type of relative position of the lines is what we might call "line relative" relations. There are two types of line relative relations. The first is called "correspondence." "Correspondence" obtains only between lines one and four; two and five and three and six and only when the content of the lines are different. Thus, the relationship of correspondence obtains between line one and line four if and only if one is a *yin* and the other is a *yang* line.

Another variety of "line relative" relations is called "holding together." "Holding together" occurs with two adjacent lines, provided they are opposite in nature. There are two aspects to this relation. The bottom line "receives" the top while the top line "rests upon" the bottom.

Although there is no exact analogy between these line relations and aspects of the macrocosm that can be supported with textual evidence, it is most probably that the notions of correctness, correspondence, centrality, and holding together were taken from observations of man and his relations in the world./51/ But the interesting point in the study of the notion of relative position in the *Great Treatise* is that not all possible relations were considered. Thus the conception of time and relative position as categories of experience rests upon wholes which are composed of hierarchy of relations./52/

I would like now to review some of the main conclusions that an examination of the arguments of the *Great Treatise* would suggest. First of all, this study does not profess to be a conclusive statement of the early Chinese notion of change. A Chinese version of *Timaeus* of Plato or the *Physics* and *Metaphysics* of Aristotle which clearly laid out these notions was never written. Moreover, the statements in the *Great Treatise* that I have used to piece these concepts together are made with reference to other arguments. Nevertheless, the technical statements about change which do occur in the *Great Treatise* provide a metaphysical grounding for the fundamental conception upon which the *Book of Changes* is based: that there is a constant but orderly and patterned development of change in the world. These technical statements, either taken as a whole or separately, in no way contradict the commentatory traditions of the earlier commentaries of the ten-wings. The *Great Treatise* merely places the *Book of Changes* within the context of competing metaphysical systems of the late Warring States period. But the larger dimensions of the organic-process conception of the *Great Treatise* always lay just below the surface of the old *I Ching*. The authors of the *Great Treatise* did not argue for their conception of change since for them it was abundantly clear and generally accepted; it was really a necessary condition for understanding the *Book of Changes* in any case. Instead they use technical statements from this conception *as evidence* for arguments they did wish to establish.

We are interested not in the derivative arguments that the authors of the *Great Treatise* wished to establish in their attempt to place the *I Ching* in a new context—arguments about their cultural history and ideas about the succession of sage kings, but in that set of presuppositions which made their arguments possible.

The evidence that we have examined tends to suggest that the modern term for change, *pien-hua*, which is seen fairly frequently in late Chou texts such as *Hsün-tzu*, *Chuang tzu*, and parts of the *Kuan tzu*, was originally a term which contained a bipolar opposition between systematic and nonsystematic change.

The term *hua*, transformation, was an important term in Taoist texts such as *Chuang tzu*. The two most important terms for systematic change, for the authors of the *Great Treatise* at least, were alternation and development, *pien* and *t'ung*. Alternation and development were generalizations from observations in nature, especially living nature. Vitalistic images such as the Creative and Receptive predominate in the *Great Treatise* as well as other of the ten-wings. Alternation of bipolar opposites and development or expansion of these opposites into an overall pattern fill the concept of time with content. The things and events which interact within this notion of time maintain rather specific relationships to each other.

These hierarchical relationships which occur within a given unit of time provide a pattern internal to that unit. Thus time and relative position become fused together; relative position is "collapsed" upon time giving time a structure and an internal pattern. In this sense the fundamental categories of experience are not time *and* relative position, but something like "time-relation."

But categories of experience and categories of explanation are merely a method of imparting order and pattern upon the flux of change. Randomness and chance were also part of the flux and defied attempts at rationality. The authors of the *Great Treatise* were careful to include this aspect of life itself into their speculation, for in their unified conception of the cosmos they wanted to express the diversity they saw in life.

NOTES

/1/ Although now more commonly known as the *Hsi tz'u chüan* than the *Ta chüan*, the former name is found only in texts of the latter Han. Cf. *Han shu* 99 *chung* 26b (SPPY ed.). The first name we have for this text is the *Ta chüan* or *Great Tradition* or *Great Treatise*. This name was given by Szu-ma Ch'ien in the *Shih-chi* 130.3a (SPPY ed.). The great Russian scholar Iu K. Shchutskii has this to say about the *Great Treatise*: "Thus, the *Hsi tz'u chüan—Tradition on the appended texts—*is an independent book which is now usually attached to the editions of the *Book of Changes*. It is divided into two parts: with regard to both its themes and its style, it is by no means a homogeneous text; it is rather a collection of short statements on the *Book of Changes* as a whole, and on various themes from the points of view of various experts of the *I-Ching*. In general, the following themes are touched upon: (1) ontology, (2) reflections on the *Book of Changes* and the adequacy of these reflections, (3) origin of the book (two interpretations), (4) embryos of epistemology, (5) ethics, (6) history of culture and the *Book of Changes*, and (7) application of the book as a mantic text and the technique of divination by it. As is obvious, this is an extremely complex work; consequently it is not surprising that it attracted the attention of scholars." Iu K. Shchutskii, *Researches on the I-Ching*, translated by William L. MacDonald and Tsuyoshi Hasegawa, edited by Hellmut Wilhelm, with an introduction by Gerald W. Swanson (Princeton, 1981), p. 84.

/2/ The first of the ten-wings is the *T'uan chüan* or *Commentary on the Decision*. This is essentially a commentary on the hexagram texts or Judgments, for each of the sixty-four hexagrams. In its structure it repeats phrases of the hexagram texts and gives a commentary on them. This commentary is couched in terms of *kang* and *jo* or "hard and soft." These terms refer to *yang* or undivided lines and *yin*, or divided lines respectively of the hexagram itself. The *t'uan chüan* is very important for an understanding of how the early Chinese viewed the structure of the various hexagrams. But the *t'uan* also occasionally used trigrams in its discussion of the hexagram structure. In this sense, it preserves two essentially discrete methods of interpretation of a hexagram: the discussion of the hexagram as a whole, and the consideration of a hexagram as a composite of two three-lined trigrams. (Cf. the *t'uan chüan* for hexagram thirty-one.) The *t'uan chüan* is divided into part A and B, and thus constitutes the first two of the ten-wings.

The third and fourth of the ten-wings is called the *Hsiang chüan* or *Commentary on the Images*. Actually, these are two completely different texts and have nothing to do with one another. The third of the ten-wings is the *Ta-hsiang chüan*, or *Tradition of the Great Images*. It utilizes the trigram method of interpretation completely and is composed of three parts: first, the trigram images or their correlations; second, the name of the hexagram which is pictured by the interplay of these two trigram images: and finally, a moral message, usually couched in terms of the acitivity of the *chün tzu,* or morally superior man.

The fourth of the ten-wings is called the *Hsiao-hsiang chüan,* or *Commentary on the Smaller Images.* It is actually a line commentary on each of the 384 lines of the sixty-four hexagrams. It repeats the line text and gives a commentary on it. The commentatory parts, taken as a whole, form a 384 line poem.

The fifth and sixth of the ten-wings is the *Great Treatise.* We shall discuss the structure of our text later in more detail.

The seventh of the ten-wings is the *Wen-yen chüan* or Commentary on the Words of the Text. This is a commentary on the first two hexagrams only, and like all the other of the ten, is traditionally attributed to Confucius. Parts of it are quoted in the *Tso chüan* and so it is probably older than the *Great Treatise.*

The eighth of the ten-wings is the *Shuo-kua chüan* or *Discussion of the Trigrams.* It is composed of ten short chapters, the first two of which are similar in style and content to the *Great Treatise.* The last eight chapters of this commentary are actually glosses and correlations to each of the eight trigrams.

The ninth of the ten-wings is the *Hsü-kua chüan* or *Sequence of the Hexagrams.* This is a series of glosses on the names of the trigrams together with a short statement showing how each hexagram relates to the preceding and the following.

The last of the ten-wings in the traditional order is the *Tsa-kua chüan* or *Miscellaneous Notes on the Hexagrams.* This is another series of glosses on the names of the hexagrams.

/3/   *Chou-i cheng-i* 7.1a (SPYY ed.).

/4/   Cf. Donald J. Munro, *The Concept of Man in Early China* (Stanford, 1969), pp. 32, 88. Joseph Needhan, *Science and Civilisation in China* (Cambridge, 1962), 2:274, 307. Needham gives two dates for the *Great Treatise.*

The tradition of Confucian authorship of the *Great Treatise* was not challenged until the tenth-century Confucian Ouyang Hsiu. In his long essay entitled "Questions of a Youth" he considers the problem of the *Great Treatise:* "A youth asked saying: 'Was not the *Hsi-Tz'u chüan* made by the sage?' I said: 'Why only the *Hsi-tz'u chüan?* The Wen-yen, Shuo-kua and the rest were also not written by the sage, but are an amassed persuasion and a muddy confusion, not even the words of one person. Those who studied the *I* in ancient times variously incorporated material for their discussions and persuasions, so that their persuasions are not even of one school Thus, some are the same; some are different; some are correct; some are wrong." *Ouyang Wen-chung ch'uan chi* 78.1a (SPPY ed.).

In modern times the opinion that Confucius did not write the *Great Treatise* has been stated by several scholars, most importantly those involved with the compilation of the *Ku shih pien,* Ku Chieh-kang, and Li Ching-ch'ih. See Ku Chieh Kang, ed., *Ku shih pien* (Discussion on Ancient History and Philosophy) (Peiping, 1916), 3:97–118 *passim.* Contemporary Chinese scholars vary on their opinions about the authorship of the *Great Treatise*, but no less an authority than Wu K'ang still suggests that the immediate disciples of Confucius, those who put together the *Analects*, were the compilers of the *Great Treatise*. See Wu K'ang *Chou-i ta-kang* (Taipei, 1957), p. 97.

/5/   *Shih chi* 130.3a (SPPY ed.).

/6/   Bernard Karlgren, "The Authenticity and Nature of the *Tso chüan*," *Göteborgs Högskolas Årsskrift*, 32, no. 3 (1926): 62–65. The *Great Treatise* conforms to Karlgren's observations on a "third century literary language."

/7/   For the notion that Fu Hsi was a local deity in Chou times see Chang Ch'i-yün, "Fu Hsi: The First Chapter of Chinese History," *Chinese Culture*, 3, no. 3. The opening paragraphs of the *wu-tu* chapter of Han-Fei tzu contains probably the longest description outside the *Great Treatise* itself of these early sovereigns. Also see the monumental study by Bernard Karlgren, "Legends and Cults in Ancient China," *Bulletin of the Museum of Far Eastern Antiquities*, 18 (1948): 1–199.

/8/   See gloss 1550 for the *hung-fan* chapter of the *Shu-ching* or Book of Documents. Bernard Karlgren, "Glosses on the Book of Documents," *Bulletin of the Museum of Far Eastern Antiquities*, 20 (1948): 245.

/9/   Alfred North Whitehead, *Science and the Modern World* (New York, 1967), p. 48.

/10/   The binome might conceivably be a combination of two early dialect words employed to expand the range of the phonetic value. A hypothetical example taken from English might be the combination of a "pail" from one dialect and "bucket" from another into one binome "bucket-pail" which would have the advantage of being intelligible to both. Hsün tzu describes the process of binomes: "Where a single name is sufficient to express the meaning, a single name should be used; where a single name is not sufficient to express the meaning, a compound name should be used. Where there is no conflict between the single name and the compound name, they may be used interchangeably to refer to the same thing as occasion demands. Although they are used interchangeably, there is no harm done." B. Watson, translator *Hsün tzu; Basic Writings* (New York, 1967), p. 143; 16.4a (SPPY ed.).

/11/   *Chou-i cheng-i* 7,16b (SPPY ed.).

/12/   *Chou-i cheng-i* 8.9a (SPPY ed.).

/13/   This usage also seems consistent with that found in Chuang tzu chapter thirty-one: *kuan tung-ching chih pien* "observe the alternations of stillness and rest." See B. Watson, translator, *Chuang tzu; Basic Writings* (New York, 1965), p. 349; 10.5a. (SPPY ed.). Also see *Hsün tzu*, chapter twenty; 14.1a (SPPY ed.).

/14/   *Chou-i cheng-i* 8.1a (SPPY ed.).

/15/   *Chou-i cheng-i* 8.11a (SPPY ed.).

/16/   For Tōdō the basic meaning of his entire series 165 is "to change the form." For *hua* Tōdō quotes the passage from the *cheng-ming* chapter of Hsün tzu which we reproduce below. On the terms *pien* and *hua* as used by Taoists, Needham has this to say:

> Concentrating their interest upon Nature as they did, it was inevitable that the Taoists should be obsessed by the problem of change. This entended to certain other schools also, especially the Naturalists and the Logicians, of whom we shall speak below. A number of technical terms were developed, such as *pien, hua, fan* and *huan* the exact meanings of which are sometimes difficult to differentiate. . . . The exact difference between *pien* and *hua* is perhaps more uncertain. In modern Chinese usage, pien tends to signify gradual change, transformation or metamorphosis; while *hua* tends to mean sudden and profound transmutation or alternation (as in a rapid chemical reaction)—but there is no very strict frontier between the words. *Pien* could be used of weather changes, insect metamorphosis, or slow personality modification of his fundamental outlook. (Needham, 2:74–75)

Also see Paul Demieville, "Résumé des Cours de l'Année Scholair; Chaire de Langue et Literature Chinoises," *Annuaire du Collège de France*, 49 (1949): 177ff. Let us look, however, at the *Kuan tzu* passage in context: "The supreme quality of Heaven is correctness, of earth equality, of man quiescence. Spring, autumn, winter, and summer are the seasons of Heaven. Mountains and river valleys are the limbs of earth. Joy and anger, taking and giving (underlie) the schemes of man. For this reason, the sage changes with the times yet is not transformed (*sheng jen yü shih pien erh pu hua*). He accords with things yet is not moved. Able to be correct and quiescent, he is thus able to remain stable. A stable heart lying within, his ears and eyes are sharp and clear, his four limbs strong and firm. And (his heart) thereby becomes the dwelling place of the essence" (W. Allyn Rickett, *Kuan-tzu* [Hong Kong: 1965], p. 160; 162,b [SPPY ed.]. The theme of *pein* and *hua* is taken up again on page 3a: "What enables transformation in unity with things is called Spirit. What enables change (*pien*) in unity with affairs is called wisdom. To transform without altering one's breath of life, To change (*pien*) without altering one's wisdom, Only the man of quality who grasps the Unity of Nature is able to do this!" (Rickett, *Kuan tzu*, p. 161; 16.3a [SPPY ed.]).

Thus it appears that in this chapter from the *Kuan tzu*, alternation is associated with the bipolar opposites of joy and anger, giving and taking in the first quotation and associated with the spiritual nature in man. To translate the problematic phrase as "the sage, in accord with the time, alternates, but does not transform," hardly captures the meaning. And yet the sense of the passage is that the sage changes in accord with the time and is not changed *by* the time. From the second quotation we get the clear idea that to be in accord with the time is wisdom. Here I feel we have a reasonable example of a technical usage of *pien* and *hua*, as defined in the *Great Treatise*.

/17/   *Chou-i cheng-i* 7.18b (SPPY ed.).

/18/ Hsün tzu gives us a beautiful description of the concept of transformation: "There are things which change their form and, although they are still the same thing in reality, appear to become something different. These are called transformed things (*hua-wu*). Although they change form, they are not distinguished anew because they are actually the same in reality" (Watson, *Hsün Tzu*, pp. 144–45; 16.4b [SPPY ed.]).

/19/ The concept of transformation is very important to Chuang tzu. It is used by Chuang tzu seventy-three times, while alternation is used forty-one times—about twice as often. An interesting subject for further research is an examination of the hypothesis that *hua* is a technical term in Chuang tzu for the alchemical transformation of the spiritual nature in man.

/20/ *Chou-i cheng-i* 7.9b (SPPY ed.). Here we see how significant statements concerning the notion of change are made en passant to a general discussion on the origins of civilization.

/21/ *Chou-i cheng-i* 8.3a (SPPY ed.).

/22/ *Chou-i cheng-i* 7.6a (SPPY ed.).

/23/ *Chou-i cheng-i* 8.6b (SPPY ed.).

/24/ *Chou-i cheng-i* 7.8b (SPPY ed.).

/25/ *Chou-i cheng-i* 7.2a and 7.3b (SPPY ed.).

/26/ There are, however, several counterexamples that do not a first glance seem to follow our formulation. The first occurrence of *pien-hua* in the *Great Treatise* is in chapter A, 1: "When images are created in heaven and forms are created on earth, *pien-hua* appear." At first sight we might be tempted to correlate *pien* with heaven or something to do with heaven and *hua* with earth or something to do with earth. In chapter A, 2, we have a similar occurrence of the two terms: "The hard and the soft displace each other and produce *pien-hua*." For this instance, the Han commentator Yü Fan explains *pien* as: "The hard displaces the soft and gives birth to *pien*. The soft displaces the hard gives birth to *hua*." Li Tao-p'ing in a subcommentary note to Yü Fan's explanation has this to say: "This is the hard and soft of the six lines. One goes and one comes, that is called *t'ui* or 'displacement,' the *yang* is called *pien* and the *yin* is called *hua*. The *yang* comes and the *yin* goes, thus the hard displaces the soft and gives birth to *pien*. The *yin* comes and the *yang* goes, thus the soft displaces the hard and gives birth to *hua*. That the hard and soft displace each other is the image of ebb and flow. Wen Wang relied on it and made nine and six (stand for) pien and hua" (cf. Li Tao-p'ing, *Chou-i chi chieh tsuan-shu* [*Ts'ung-shu chi-ch'eng*] 469 [chüan 9], p. 369).

We can see from this quotation from Yü and Li that from Han times on the traditional interpretation correlated *pien* as a kind of *yang* change and *hua* as a kind of *yin* change. This type of thinking certainly seems reasonable with respect to discussions of polarity and analogy. The quotations involved all contain bipolar terms: "hard and soft"; advance and retreat. The obvious implication is that *pien* and *hua* are also bipolar terms. Furthermore, by implication of analogy used so often in the *Great Treatise*, it follows that *pien* has some relationship by analogy to the term "hard" and the term "advance" and *hua* has a similar relation to the terms "soft" and "retreat." Moreover, since softness and retreat are aspects of the *yin* principle in nature and hardness and advance are aspects of the *yang* principle, it seems logical to conclude that Yü Fan's and Li Tao-p'ing's explanation is substantially correct. And were it not for several other quotations in the *Great Treatise* employing the terms *pien* and *hua*, the matter would be allowed to rest.

What is misleading in these quotations is that the analogy is *not* between the terms heaven and alternation and earth and transformation, or between the terms hard and alternation and soft and transformation. There is analogy at work, but we should interpret the statement as: "The hard and soft displace each other and produce alternation—and the remainder is called transformation."

/27/   For *t'ung* Karlgren, GSR 1185r, cites a number of pre-Han texts with different usages: "penetrate, pass through, communicate with, have relations with, reaching everywhere, universal, all." Tōdō is "to pierce, to penetrate, to run or stick into." My translation of *t'ung* as development is a contextual interpretation for which there is no precisely adequate English definition. Dr. Frederick Mote in personal communication has further reservations about the translation of *t'ung* as development: "It seems to me that 'development' to the average English reader stresses a value commitment to the end of the development, to the developed. I do not feel that the Chinese includes that element of meaning. I would prefer a translation that stresses the idea of 'unblocked, free, mobile,'—and therefore have tried to use 'process' or 'in process' instead of 'developing' or 'development' in all the places you use the term. I realize that this may have unsatisfactory Whiteheadian overtones, but I find that, or some other such term, better than 'development.' "

/28/   *Chou-i cheng-i* 7.18b–19a (SPPY ed.).

/29/   *Chou-i cheng-i* 7.16b (SPPY ed.).

/30/   Thus, Feng Yu-lan, *A History of Chinese Philosophy* (Princeton, 1952) 1:380, suggests that *i* means "easy": "Perhaps this explains the *I Ching's* alternative name of *Chou-i*. It was named chou from the fact that it was composed by the people of the Chou dynasty, and *i* because its method of divination was an easy one." James Legge followed Yü Fan, without acknowledgement, that the character *i* is a combination of two characters, *jih* and *yüeh* or sun and moon. Richard Wilhelm did not discuss the problem. Dr. Hellmut Wilhelm initially followed Hsü Shen's theory that the character was a pictograph of a lizard, the conceptual significance of which was derived either from color changes or rapid change of location. In 1957, following Erwin Reifler, he suggested that the concept was derived from the concept of the fixed and the straight. Unfortunately Reifler's study was never published. See James Legge, trans., The *I Ching* (New York, 1963), p. 38, and Hellmut Wilhelm, "The Concept of Time in the *Book of Changes*," *Man and Time* (Papers from the Eranos Yearbooks 3; New York and London, 1957), p. 212, n. 2.

/31/   Karlgren, *Documents*, p. 33, translation altered. Karlgren takes *i* here in the sense of "a failing" or "a shortcoming."

/32/   *Chou-i cheng-i* 7.9b and 7.18b (SPPY ed.).

/33/   The Chinese philosophers of the later Warring States period generally accepted the notion that the world was in a constant flux, and they attempted to develop a rational system of interrelated patterns which maintained some semblance of stability within the flux. The *Book of Changes* is this sort of attempt. The *Great Treatise* represents the fusion of later *yin-yang* speculation with a very old divination system. The authors of the *Great Treatise* went further with respect to the more fundamental notions of the ultimate categories of experience than did Tsou Yen and his school of naturalists.

/34/   *Chou-i cheng-i* 7.9a.

/35/   *Chou-i cheng-i* 7.17a.

/36/   I do not suggest that the development of the twenty-four solar terms had anything to do with the patterns found in the hexagrams. It is only that both are results of the same kind of preoccupation: an interest in finding orderly progression in nature. (Needham, 3:404, acknowledges their early date. Derk Bodde describes this concept of change as "The belief that the universe is in a constant state of flux but that this flux follows a fixed and therefore predictable pattern consisting either of eternal oscillation between two poles or of cyclical movement within a closed circuit; in either case the change involved is relative rather than absolute, since all movement serves in the end only to bring the process back to its starting point" (Derk Bodde, "Harmony and Conflict in Chinese Philosophy" *Studies in Chinese Thought*, ed. Arthur F. Wright [Chicago, 1951]), p. 21.

/37/   See Mircea Eliade, *Cosmos and History* (New York, 1959), especially chapter 2, for an account of certain aspects of cyclical time. Time is also a principle of organization in Jung's theory of synchronicity. See "Synchronicity: An Acausal Connecting Principle," *The Structure and Dynamics of the Psyche* (Collected Works of C. G. Jung, vol. 8).

/38/   *Chou-i cheng-i* 8.1b (SPPY ed.).

/39/   The circular model manifests completeness in the sense that the universe of discourse of the four seasons admit of no further elements—no fifth season, for example. There is also a richness of relations in that on the circular model each season is connected with each of the others and also more or less equally, since an imbalance in one yields consequences for all the others. Speaking about the Chinese sixty-year cycle of dating, Derk Bodde has this to say: "These chronological techniques are perhaps symptomatic of a Chinese world view which sees the human world as a part of the universal macrocosm and hence as conforming, like the latter, to an inherent pattern of cyclical rather than linear movement" (Derk Bodde, "Harmony and conflict in Chinese Philosophy," in *Studies in Chinese Thought*, p. 27).

/40/   Dr. Hellmut Wilhelm's pioneering article in the *Eranos Jahrbuch* series is the only thing we have on this fascinating subject. See Hellmut Wilhelm,

"The Concept of Time in the *Book of Changes*," *Papers from the Eranos Jahrbuch*, vol. 17.

/41/   Richard Wilhelm, *The I Ching or Book of Changes*, trans. from the German by Cary Baynes, 3d ed., with a preface by Hellmut Wilhelm (Princeton, 1967), p. 371. Hereafter Wilhelm/Baynes.

/42/   Wilhelm/Baynes, *The I Ching*, p. 378.

/43/   It appears that the conception of time at work in the *Great Treatise* more clearly resembles what we in the West call "psychological time" or "relative time," rather than something similar to a Newtonian or Kantian "absolute time."

/44/   The problem of space is handled in the *Great Treatise* with the formula XY *chih chien*, "the space between X and Y." Most often X and Y are heaven and earth, as in the phrase "between heaven and earth." It has been suggested to me by my colleague Luther H. Martin, Jr., that relative position is itself a spatial metaphor, but one in which the relational aspects are emphasized. This is different from the Aristotelian concept of space as extension.

/45/   *Chou-i cheng-i* 7.1a (SPPY ed.).

/46/   In chapter A, 7, we have another important reference to relative position: "Heaven and earth establish positions and change moves in their midsts." A similar statement is found in Hsün tzu, chapter 26: "Heaven and earth change position" (18.10a [SPPY ed.]).

/47/   Watson, *Chuang tzu*, p. 146:5.114b (SPPY ed.).

/48/   This is a surprising citation in view of the argument against such divisions in chapter 2. Perhaps this chapter was written by a later disciple and represents a more eclectic view?

/49/   We find this technique of derivation of larger numbers of categories of explanation from either one primary category or, in this case, two primary categories, elsewhere in the *Great Treatise*. For example, in chapter A, 21, we have: "For this reason the Changes includes the great ultimate that generates two principles. The two primary principles generate the four images. The four images generate the eight trigrams."

/50/   But the most important source for the notion of relative position within the lines of the hexagram occurs in the *t'uan chüan*, or Commentary on the Decision, and the *hsiao-hsiang chüan*, or Commentary on the Images. The term *wei* occurs seventeen times in the *t'uan chüan* and thirty-five times in the *hsiao-hsiang chüan*. In The *hsiao-hsiang chüan* the most common formula is *pu-tang wei*, "the place is not appropriate." This is often an explanation for the occurrence of the term "misfortune" in the line texts. This formula occurs sixteen times and fifteen of these instances refer to lines three or four. It is to this kind of observation that the *Great Treatise* explains line four as mostly feared and line three as mostly unlucky.

/51/   The Chinese term for correctness is *cheng*; for correspondence, *ying*; for centrality, *chung*. "Holding together" and "receiving" are not accorded a separate technical term, but the *t'uan chüan* for hexagram 59 does use the term *te* in the sense of receiving. See Wilhelm/Baynes, 690, under Commentary on the Decision.

/52/   A thoroughgoing organic view of the universe might well treat all relations equally. But for the authors of the *Great Treatise*, of all possibilities, only certain specific relations were singled out as important. This selection of significance was not random or arbitrary, but betrayed the Chinese notion that only things of the same kind influence each other. For Needham's criticism of Levy-Bruhl on this matter see Needham, 2:28–46 and 291 ff.

CHINESE CHARACTERS
(Listed in order as they occur in the text)

大傳

易經

象

小象

文言

大象傳

説卦

序卦

杂卦

道德經

史記

天地

天道

天雲

呂氏春秋

戰國策

荀子

韩非子

庄子

管子

左傳

象傳

變化

孟子

陰陽

坤

乾

通

洪範

尚書

管子

剛和柔

觀動靜之變

聖人於事變而化物

退

正

應

忠

Ancient Chinese Cosmology and *Fa-chia* Theory

Vitaly A. Rubin

Although the cosmological concepts of Yin-Yang and Five Elements appeared in China very early—they are mentioned in the *Shu ching*—the most important discussions in the first philosophical works concentrated not on cosmology but on the questions of ethics, politics, and the education of a noble character. In *Analects, Mencius,* and *Hsün Tzu* the problems of cosmology do not play a major role. Only in the second century B.C., in Tung Chung-shu's system, do we see an ideological synthesis of cosmology with the "New Text School" tradition of commentaries on the Confucian classics./1/ From that time, cosmological speculation began to be viewed in conjunction with Confucianism; furthermore, this connection was projected back to the pre-Ch'in era./2/

There is evidence, however, that Tung Chung-shu was not the first thinker who tried to create a synthesis of political theory with cosmology. In the time of Ch'in Shih-huang, the state ideology was a mixture of the ideas of *Fa-chia* theory and the cosmological considerations of the school of Tsou Yen (ca. 350–270 B.C.). In chapter 6 of the *Records of the Historian,* Ssu-ma Ch'ien describes this mixture:

> Ch'in Shih-huang advanced the teaching of the succession of Five Powers *(wu te)*. He considered Chou to have obtained the Power of Fire, and Chou had been replaced by Ch'in; therefore, he proclaimed his power to be the one that Chou Power could not overcome. Thus the Power of Water had now begun.
>
> He changed the beginning of the year and the felicitations (on this occasion) at court. Everything had to start from the first day of the tenth month. The black color was honored in all gowns, coats, scepters and banners. Six had to be the standard in numbers: plates for contracts, and official hats were six inches large: the carriages were six feet long; six feet equalled one yard; and a horse team consisted of six. The name of the Yellow River was changed to "Powerful Water" because he considered this era to be the beginning of the Power of Water. With harshness, violence, and extreme severity, every matter was decided by law; for by punishing and oppressing, by having neither humanity,

nor kindness, moderation, or righteousness, there would come
an accord with the norms of Five powers. Then they zealously
applied the laws and for a long time no one was shown any
mercy./3/

Up to now, Ch'in ideology has been interpreted as simply a continuation of *Fa-chia* political theory./4/ But this remarkable text (hereafter
referred to as "T") indicates that the ideology comprised a definite cosmological worldview as well as a political doctrine. In this article I shall
try to determine the content of this worldview, and account for its inclu-
sion in Ch'in state ideology./5/

Tsou Yen's theory of Five Powers (*wu te*) is a combination of the
conception of Five Elements (*wu hsing*) which he had taken from the
treatise "Hung Fan"/6/ and which can be understood as a Chinese vari-
ant of primitive classifications, with the yin-yang conception of pulsating
changes of darkness and light in the movement of time./7/ On the basis
of these two conceptions, Tsou Yen put forth the idea of the movement
of Five Powers in time. This movement can be seen in two cycles: the
great historical cycle of dynasties and the small natural cycle of the
year./8/

The Historical Cycle of Dynasties

As we have seen in T, Ch'in Shih-huang identified his victory over
Chou with the victory of the Power of water over the Power of fire. The
source of such an identification is Tsou Yen's idea that "each of the Five
Powers is followed by one that it cannot overcome. The dynasty of Shun
ruled by the Power of Earth; the Hsia dynasty ruled by the Power of
Wood; the Shang dynasty ruled by the Power of Metal; the Chou
dynasty ruled by the Power of Fire."/9/ Such an order of succession
received in sinological literature the title of "mutual conquest
order."/10/

The change of dynasties is thus seen as a natural process integral to
the cycle of changes of Powers that are embodied in these dynasties. The
relationship between these powers is one of struggle, with victory to the
strongest. The concept of victory plays a crucial role in the cyclical progression of powers from earth to wood, metal, fire, and water: victory of
one power over another is apparently associated with victory of its pri-
mal element in life's daily situations. For instance, the victory of metal
over wood was seen in the image of an axe that fells a tree; the victory
of fire over metal was represented by the melting of metal, etc./11/

Such an approach was in full consonance with the spirit of the Warring States period in which the division of China into several independent states was seen as a transition from the domination of the Chou

dynasty to the domination of some other future dynasty. The themes of historical change, of struggle and victory, were the central themes for the thinkers of this period. Confucians as well as legalists were concerned with determining the appropriate means and methods to effect the unification of China. Mencius (1 A 6) recommended humanity and care for the people as the best method because people under heaven would side with a human ruler;/12/ Shang Yang insisted that the right way to unite China was by force./13/

From the modern point of view the opinions of Shang Yang may seem more realistic. Indeed, one must realize that Mencius was following an ancient tradition, the beginning of which can be discerned in the concept of the Mandate of Heaven. Mencius's view may be considered as a development of Chou-kung's idea that heaven gives its mandate only to the virtuous ruler who fulfills his religious obligations and protects the people./14/ These concepts are a part of an integrated ethico-religious worldview whose roots lie in the hallowed past.

In contradistinction, Shang Yang's theory is devoid of any connection with religious or philosophical traditions. Han Fei Tzu believed that such a lack constituted a dangerous weakness in legalist political theory, and tried to remedy the situation by introducing a Taoist-legalist ideological amalgam./15/ As for Shang Yang himself, the *Book of Lord Shang* contains an attempt to present a legalist philosophy of history in opposition to the Confucian ethical approach. In chapter 7 a theory of three epochs is presented:

"In the highest antiquity people loved their relatives and were fond of what was their own; in middle antiquity, they honoured talent and talked of moral virtue, and, in later days, they prized honour and respected office."/16/ But this attempt to depict Confucianism as a teaching that belonged to the past, and was therefore unfit for the present, could not remove the ideological difficulties of Legalists.

It was Tsou Yen's theory that provided the Legalist political doctrine wth a cosmological and philosophical foundation. Based on a respected ancient treatise, "Hung Fan," which forms a part of the classical *Book of History* (*Shu ching*), proponents of the Five Powers theory introduced an entirely new interpretation of historical change: one dynasty does not replace another as a result of the moral degeneration of the old one and the virtuous behavior of the founder of the new dynasty; rather, the replacement occurs in the course of the natural and inevitable cyclical process of struggle (which was understood by Legalists as war) and victory. For the proponents of the Five Powers theory, as well as for legalists, historical change had no connection at all with moral or religious values. Although the followers of Tsou Yen made use of the concept of heaven, it was for them—in contradistinction to Mencius—not a benevolent deity caring for people, but a natural principle determining the orderly succession of Powers.

The proponents of Tsou Yen's theory understood the role of the politician in a manner entirely different from Confucians: if for the latter the central purpose had consisted in helping and educating the people, for cosmologists the crucial part of political art was the ruler's ability to understand the timing of dynastic change and to align himself with the power of the future by introducing measures that bring the state ceremonies and calendar into accord with this power. Such a task demands that a politician recognize the portents presaging the epochal change. In a passage in "Lü-shih Ch'un Ch'iu" there is a description of the events marking a change of dynasties:

> Whenever the emperor or king is about to arise, Heaven must first send some favorable omen to the lower people. In the time of Yellow Emperor, Heaven first made a large number of earthworms and mole crickets appear. The Yellow Emperor said: "The ch'i/17/ of Earth is winning." Therefore he assumed yellow as his color, and took Earth as a pattern for his affairs.

> In the time of Yü, Heaven first made grass and trees appear which did not die in the autumn and winter. Yü said: "the ch'i of Wood is winning." Therefore he assumed green as his color, and took Wood as a pattern for his affairs.

> In the time of T'ang, Heaven first made some knife blades appear in the water. T'ang said: "The ch'i of Metal is winning." Therefore he took white as his color, and took Metal as a pattern for his affairs.

> In the time of King Wen, Heaven first made a flame appear, while a red raven holding a red book in his mouth alighted on the altar of the soil of Chou dynasty. King Wen said: "The ch'i of Fire is winning." Therefore he assumed red as his color and took Fire as a pattern for his affairs.

> Water will inevitably replace Fire, and Heaven will make the victory of Water's ch'i manifest. Victory of the ch'i of Water means that black has to be assumed as its color, and Water must be taken as a pattern for affairs. If the ch'i of Water arrives without being recognized, the operation, when its cycle is complete, will revert once more to Earth./18/

The last phrase emphasizes the significance of recognizing the signs; this is possible only if one understands the correlation between events in nature and in the human world. Tsou Yen and his followers were specialists in these correlations, which explains the fact of their unusual popularity with the rulers of the time./19/

The Natural Cycle of the Year

In Tsou Yen's theory the succession of Five Powers not only exists in the movement of history, but also in the cycle of the seasons of the year. If, in the conception of the historical movement, attention is paid not so much to the powers themselves but to the moments of their change, here, in the natural cycle, the central role is played by the qualities of powers connected with the character of seasons. Spring, for instance, being the time of plants' growth, is an embodiment of the Power of Wood; summer, a hot season, is an embodiment of the Power of Fire. Accordingly, this order of succession—the so-called "mutual production order"/20/—is different from the "mutual conquest order"; here Wood is followed by Fire, Earth, Metal and Water./21/

The idea of correlations permeates the year's cycle in even greater measure than the historical one. Already in "Hung Fan," in which the conception of Five Elements (later transformed by Tsou Yen to Five Powers) is first expounded, one can see a number of correlations between the world of nature and the world of man: e.g., the element Water corresponds, among tastes, to salt; among virtues, to gravity; among vices, to wildness; among favorable verifications, to timely rain./22/ This feature of the Five Elements concept is apparently connected with the role played by divination and magical practices in archaic Chinese thought, as in every system of primitive classification./23/

The emphasis on correlations is also a major characteristic of Chinese calendars composed under the influence of Tsou Yen's ideas. In "Yeh ling," correspondences are established between the season and Power, particular number, taste, color, kind of sacrifice, etc./24/ There are the same correlations in the calendar "Yu kuan."/25/ Ch'in Shih-huang's measures, mentioned in T, are in fact a practical realization of instructions contained in calendars about the correlation of Water with winter, the number six, and the color black. As for renaming the Yellow River "Powerful Water," this measure, similar to the frequent renamings in communist countries, is typical of the absurdities carried out by despotic rulers devoted to a particular ideology.

The Rehabilitation of Tsou Yen by Ssu-ma Ch'ien

The last phrase of T suggests that the theory of Five Powers entailed rejection of moral values and advocacy of a repressive policy. There is an analogous indication in chapter 28 of the "Records of the Historian," in which Ssu-ma Ch'ien writes that after Ch'in Shih-huang had decided to identify with the Power of Water, "in all his government affairs he put laws before everything else."/26/

I have noted the features of Tsou Yen's theory that made it possible for legalists to use it as a philosophical foundation for their political doctrine. It is necessary, however, to point out the marked difference between the attitude of value neutrality typical of Tsou Yen's teaching, and the legalist tendency to create an emphatically inhuman state machine to implement a system of rewards and punishments./27/ There is good reason to suppose that Ch'in Shih-huang was interested in concealing this difference, and representing his policy as a realization of cosmic law. Such a tactic would dishearten his enemies, making them feel that their struggle against his domination was tantamount to a struggle against elemental forces. This is why I suggest that the last phrases of T may be a citation or exposition of some lost document written by Ch'in Shih-huang or his henchmen. As I will show later, it could not be the opinion of Ssu-ma Ch'ien because he tried to prove the opposite thesis according to which Tsou Yen was a Confucian.

Attentive reading of calendars gives a hint about the likely pretext used by Ch'in Shih-huang in concocting his ideological falsification. Winter, in its correlation with the Power of Water, was considered to be a time of severity. In "Yu kuan," in the part related to winter, we find an entry: "During the twelve days of the first cold complete your punishments."/28/ Tung Chung-shu apparently was influenced by Tsou Yen's ideas when he wrote that "beneficence, rewards, punishments and executions match spring, summer, autumn and winter."/29/ These declarations, however, are formulated in general terms, and placed among other declarations which sometimes contradict them. For instance, immediately after the entry from "Yu kuan" cited above there is an instruction to "bestow favors during the next twelve days." A description of winter activities in the same calendar includes the statement that when the "prince utilizes number six, his temper is compassionate and kind."/30/ In calendar "Yüeh ling," "severity" finds expression only in relation to people who "encroach on or rob the others."/31/

There are a number of texts that show the authority of the Five Powers theory in third-second century B.C. to be too entrenched to suffer any attenuation from the fall of the Ch'in dynasty. Ssu-ma Ch'ien relates that the first Han emperor, Kao-tsu, who had abolished the laws of Ch'in,/32/ "continued the old Ch'in practice of beginning the new year with the tenth month and made no change."/33/ He also describes a discussion that occured in the second century B.C., between two learned men, about the question of which Power corresponded to the Han dynasty. After the appearance of Yellow Dragon it was considered proven that Han Power was a Power of Earth; one of the scholars received instructions to draft a new calendar and set of regulations based upon the theory of the domination of the Power of Earth./34/

At the same time, Ssu-ma Ch'ien apparently felt that Ch'in Shih-huang's usage of the Five Powers theory as justification for his own repressive policy casts a shadow on its creator Tsou Yen. In Ssu-ma Ch'ien's time, some of Tsou Yen's ideas were included by Tung Chung-shu in his ideological synthesis/35/ and, therefore, became a part of official Confucianism. In this light the question of Tsou Yen's Confucian credentials was linked to the problem of the purity of the sources of official Confucian doctrine. This seems to explain Ssu-ma Ch'ien's efforts to rehabilitate Tsou Yen./36/

The biography of Tsou Yen in chapter 74 of the "Records of the Historian" is placed between the biographies of the most prominent Confucian thinkers, Mencius and Hsün Tzu./37/ It begins with the following declaration: "Tsou Yen saw that the rulers were becoming ever more dissolute and were not disposed to exalt virtue." Therefore, as Ssu-ma Ch'ien writes, Tsou Yen "scrutinized the operations of Yin and Yang . . . and wrote essays totalling more than one hundred thousand words about the appearance of amazing changes and of the rise and fall of great sages. His words were grandiose and unorthodox. He invariably began by examining small objects and extended the examination to larger and larger ones till infinity." Setting forth Tsou Yen's geographical concepts, Ssu-ma Ch'ien writes: "His theories are all of this sort. But in the final analysis they all come back to the virtues of humanity (*jen*), righteousness (*i*), moderation (*tzu*) and frugality (*chien*), and the observance of proper relations between ruler and subject, the superior and the inferior, and the six family relationships. It is only at the start that his theories are so extravagant."

But it seemed to Ssu-ma Ch'ien that such a defense was insufficient. The historian considered it necessary to refer not only to Tsou Yen's ideas but also to his conduct. The honors with which the ancient Chinese rulers cloaked the cosmologist were in striking contrast to the persecutions which befell Confucius and Mencius. Noting this contrast, Ssu-ma Ch'ien makes it plain that the aims of Tsou Yen were no different from those of the founding fathers of Confucianism but, unlike them, he acted by indirect methods. The model advisers, I Yin and Pai-li Hsi, often cited by Confucians, also resorted to such methods. Just as I Yin became a cook in order to win the trust of Ch'eng T'ang, and Pai-li Hsi tended oxen in order to get close to Ch'in ruler Mu-kung, so did Tsou Yen elaborate his cosmological theories to ingratiate himself with the rulers. Once he acquired their trust, he could lead them onto the path of the great Tao.

Neither of these attempts to depict Tsou Yen as a Confucian can stand up to criticism; both his concepts and his conduct bore no relation to Confucianism. But, to a certain degree, Ssu-ma Ch'ien achieved his aim. He bolstered the tradition which linked the pre-Ch'in cosmology to Confucianism in spite of the fact of its proximity to some features of *Fa-chia* theory.

NOTES

/1/ Some sinologists view the Confucianism of Tung Chung-shu as sharply distinct from classical Confucian thought; Wing-tsit Chan, for example, refers to Tung's views as "Yin-Yang Confucianism." See his *Source Book in Chinese Philosophy* (Princeton, 1969), 271.

/2/ Fung Yu-lan, *A History of Chinese Philosophy* (Princeton, 1953), vol. II, 9.

/3/ See Ssu-ma Ch'ien, *Shih Chi*, Shang wu yin shu kian, 112; Edouard Chavannes, *Les Mémoires Historiques de Se-ma Ts'ien* (Paris, 1897), vol. II, 130; Derk Bodde, *China's First Unifier* (Leiden, 1938), 113.

/4/ See Bodde, *China's First Unifier*, 181–222; L. S. Perelomov, *Kniga pravitelia oblasti Shang*, (Moscow, 1968), 116–20; Vitaly A. Rubin, *Individual and State in Ancient China*, translated by S. Levine (New York, 1976), 86.

/5/ The most detailed treatment of the problem of Ch'in ideology can be found in *China's First Unifier*; Bodde cites T (113–14), but does not touch on the question of the relationship between *Fa-chia* and Five Powers theories because, in his opinion, the Five Powers theory influenced Ch'in Shih Huang, but not Li Ssu, the chief architect of Ch'in successes and a champion of *Fa-chia* doctrine.

/6/ James Legge, translator, *The Chinese Classics* (Hong Kong, 1960), vol. III, 320–44.

/7/ Vitaly A. Rubin, "The concepts of Wu-hsing and Yin-yang," forthcoming.

/8/ In the second volume of his *Science and Civilisation in China* (Cambridge, 1956), Joseph Needham says "there is considerable reason to suspect that [the Tsou Yen school's] 'arts' included . . . calendrical science" (239). Needham's caution here may be explained by the fact that at the time he wrote those lines, this problem had not been adequately investigated. With the publication of W. Allyn Rickett's *Kuan Tzu* (Hong Kong, 1964), however, no doubt can exist about the influence of Tsou Yen's school on calendars. See especially Rickett's chapter focusing on "Yu kuan," 180–219.

/9/ Needham, *Science and Civilisation in China*, II, 238.

/10/ *Ibid.*, 256.

/11/ See Bodde, *China's First Unifier*, 113.

/12/ James Legge, *The Works of Mencius* (New York, 1970), 136.

/13/ J. J. L. Duyvendak, *The Book of Lord Shang* (Chicago, 1963), 190.

/14/ Legge, *The Chinese Classics*, III, 431.

/15/ According to Han Fei Tzu's biography in chapter sixty-three of the *Shih Chi*, Ch'in Shih Huang was enthusiastic about the ideas of Han Fei. It is easy to understand, however, why Ch'in Shih Huang preferred the union of *Fa-chia* with the Five Powers theory rather than Taoism. The individualistic and anarchistic Taoism of "Chuang Tzu" could not have any connection with the Ch'in official doctrine. Taoism, as expounded in the *Lao Tzu*, was not deemed fit to be part of the ideology of a totalitarian empire, mainly because Lao Tzu prescribed nonaction for the ruler, i.e., noninterference in the lives of others, and the provision of opportunity for the spontaneous development of every individual (see especially ch. 57). This contradicts the aims of legalist rulers, and the means used by these rulers to achieve their goals—violence and war—are censured in the *Lao Tzu* repeatedly.

/16/ Duyvendak, *The Book of Lord Shang,* 226.

/17/ The meaning of *ch'i* here—"breath," "ether," "matter-force"—appears close to the meaning of *hsing*—"element," "agent," "force." Its appearance in this text shows that the Five Powers theory is rooted in the Five Elements classification.

/18/ Fung, *A History of Chinese Philosophy*, I, 161.

/19/ *Ibid.*

/20/ Needham, *Science and Civilisation in China*, II, 255.

/21/ James Legge, translator, *Li Chi* (New York, 1967), I, especially 249–310. See also Rickett, *Kuan Tzu*, 198–219. The discrepancy created by four seasons and five forces was resolved by the introduction, in calendar "Yüeh ling," of a part consecrated to Power Earth in the midst of the four seasons, between summer and autumn. In the calendar "Yu kuan" this difficulty was resolved by arranging the seasons on a chart with Power Earth forming a square in the center, and the four seasons arranged around it. See Rickett, 182–93.

/22/ Legge, *The Chinese Classics*, III, 327–36.

/23/ Claude Lévi-Strauss, *The Savage Mind* (London, 1976), 43.

/24/ Legge, *Li Chi*, I, 249–54.

/25/ Rickett, *Kuan Tzu*, 199–219.

/26/ Ssu-ma Ch'ien, *Shih Chi*, 736.

/27/ Vitaly A. Rubin, "Shen Tao and Fa-Chia," *Journal of the American Oriental Society*, 94,345.

/28/ Rickett, *Kuan Tzu*, 209.

/29/ Fung, *A History of Chinese Philosophy*, II, 48.

/30/ Rickett, *Kuan Tzu*, 210.

/31/ Legge, *Li Chi*, I, 301–04.

/32/   Ssu-ma Ch'ien, *Shih Chi*, 183.

/33/   *Ibid.*, 1605.

/34/   *Ibid.*, 1606.

/35/   See note 8.

/36/   Needham also sees the declarations about the Confucian character of Tsou Yen's doctrine as Ssu-ma Ch'ien's efforts to rehabilitate Tsou Yen (*Science and Civilisation in China*, II, 235). But he interprets these efforts as necessary to account for the fact that Confucius had been a failure from the worldly point of view; this shows a lack of understanding of the Confucian theory of the *chün tzu*, who would be ashamed to be rich and honored when there is no Tao under heaven (*Analects*, 8.13).

/37/   Ssu-ma Ch'ien, *Shih Chi*, 1347–49.

CHINESE CHARACTERS

(1) *ch'i*   氣

(2) *Ch'in Shih Huang Ti*   秦始皇帝

(3) *Fa Chia*   法家.

(4) *Kuan Tzu*   管子

(5) *Shih Chi*   史記

(6) *Ssu-ma Ch'ien*   司馬遷

(7) *wu hsing*   五行

(8) *wu te*   五德

Concepts of Comprehensiveness and Historical Change in the *Lü-shih ch'un-ch'iu*

John Louton

Among the commentators' remarks and bibliographic entries, there are none that express any doubt concerning the manner in which the *Lü-shih ch'un-ch'iu* 呂氏春秋 was composed./1/ All are agreed the text is the work of many scholars working under the patronage of Lü Pu-wei 呂不韋 . Not only is there general agreement concerning the multiplicity of scholars who contributed to the composition of the *Lscc*, but also to the contention that these scholars represented many, if not all the diverse philosophic schools of the day. Clearly, implicit in any argument that the *Lscc* maintains a reasoned and consistent position on any issue is an intimation that the *Lscc* is a truly eclectic philosophical text, as opposed to merely an encyclopedia of late Chou philosophic speculations. This paper does not pretend to make a definitive statement on this issue, since to do so would require a thorough study of the consistency of the positions maintained in the text on every major issue it treats. However, since the subject to be dealt with in this paper is the consistent nature of the position maintained in the *Lscc* regarding the notions of "historical change" and "comprehensiveness," one could certainly consider this paper to be a step in the direction of showing the *Lscc* to be a truly eclectic philosophical text.

The thesis I will attempt to demonstrate in this paper is this: the authors of the *Lscc* view historical change as the product of an interaction between an unaltering sequence of natural phenomena and concurrent human events. The text is unclear as to the exact nature of the influence each of these has on the other, but the basic view is that if those in control of the course of human events, i.e., rulers, are capable of understanding the sequence of natural phenomena and guide the course of human events into a proper correspondence with it, the results will be beneficial; if not, the results will be disastrous. Up to the present, there has been a steady decline in the ability of those in power to understand either the important influence of the sequence of natural events, or the significance of the course of past human events. As a result, there has been a steady decline in the historical process (that is, the past is viewed as clearly superior to the present) since the early days of the sage-kings. This decline, however, is not viewed as inexorable. On the contrary, it can be curtailed, but only if those in power gain a comprehensiveness of knowledge which would allow them to discern the order of the sequence of natural phenomena, their present position in it, and the true significance of past human events. With a complete understanding of these three things, according to the text, one knows which natural influence is in power, which will be next, and what the proper response to both the current and ensuing natural influence is. As a result, one can possess foreknowledge of changes to come and can prepare a proper response.

Clearly, in order to demonstrate the validity of this statement of the *Lscc*'s position on historical change and comprehensiveness four things must be established: (1) the text espouses a belief in an unaltering sequence of natural phenomena; (2) that sequence does influence, and is influenced by, the course of human events; (3) there has been a steady decline in the historical process manifested in the superiority of the past to the present; (4) the only way to reverse the decline in the state of the human condition is for a ruler to gain a comprehensiveness of knowledge which includes an understanding of the sequence of natural phenomena, an awareness of one's present position in that sequence, and a knowledge of the true significance of past events.

Demonstrating the first of these presents little difficulty. The *Lscc* repeatedly espouses the *wu-hsing* 五 行 explanation of natural phenomena. For example, one *p'ien* 篇 entitled "Ying-t'ung" 應同 says,

> Whenever any emperor or King is about to arise, Heaven must first make manifest some favorable omen among the lower people. In the time of the Yellow Emperor, Heaven first made a large (number of) earthworms and mole crickets appear. The Yellow Emperor said: 'The force of the element earth is in ascendancy.' Therefore he assumed yellow as his color, and took earth as a pattern for his affairs.
>
> In the time of Yü, Heaven first made grass and trees appear which did not die in the autumn and winter. Yü said: 'The force of the element wood is in ascendancy.' Therefore he assumed green as his color, and took wood as a pattern for his affairs.
>
> In the time of T'ang, Heaven first made some knife blades appear in the water. T'ang said: 'The force of the element metal is in ascendancy.' Therefore he assumed white as his color, and took metal as a pattern for his affairs.
>
> In the time of King Wen, Heaven first made a flame appear, while a red bird, holding a red book in its mouth, alighted on the altar of soil of the House of Chou. King Wen said: 'The force of the element fire is in ascendancy.' Therefore he assumed red as his color, and took fire as a pattern for his affairs.

Water will inevitably be the next thing which will replace
fire. And Heaven will first make the ascendancy of water manifest. The force of water being in the ascendancy, black will be
assumed as its color, and water will be taken as a pattern for
affairs. If the power of water arrives without being recognized,
the operation, when its cycle is complete, will revert once more
to earth./2/

Here we have a clear statement of the view that there is a recurring
sequence of "natural forces" which have an influence on the course of
human events. This passage continues:

What Heaven makes are the seasons. And yet, it does not
help complete agricultural labors, below. Things of a similar sort
summon each other. When *ch'i* 氣 is the same, then there is
agreement. When tones are similar, then there is response. Strike
a *kung* 宮 note, and a *kung* note will activate. If water is poured
out on flat earth, it will flow to damp places. If one spreads kindling equally around a fire, it will burn the kindling that is dry.
Mountain clouds are like grasses and scrubs. River clouds are like
fish and scaly things. Arid clouds are like smoke and fire. Rain
clouds are like water and waves./3/ There is nothing that does
not resemble that which produced it; thus it is typified to man.
Therefore, the dragon brings about rain, and form is followed by
shadow, and where an army is placed, so certainly must thorns
and brambles grow.

Mankind regards fate as a source of good and bad fortune.
How can they know its source. If one tips over a nest and
destroys the eggs, the phoenix will not come. If one cuts open
animals and eats their fetuses, the unicorn will not come. If one
drains sloughs and dries up the fishing, dragons and turtles will
not remain. All the instances of things following what is similar to
them cannot be recorded.

A son cannot be impeded by his parents. A minister cannot
be impeded by his ruler. When there is harmony with the ruler,
he comes along. When there is a difference with the ruler, he
goes away. Therefore, the ruler, though he is honored, if he
regards white as black, the minister cannot listen. The father,
though he is a parent, if he regards black as white, the son cannot
follow. . . .

An exhortation of the Shang says, 'Heaven's sending down
calamities and manifesting auspicious omens both have a function. They are used to explain that bad and good fortunes are
sometimes brought about by men themselves.' Therefore, when a
state is disordered, it is not only disordered, but must also bring
on bandits. If it were only disordered it would still not necessarily
be destroyed. Yet, if it brings on bandits, then there is no way for
it to be preserved.

> With regard to the uses of the military, it can be used either
> for profit or righteous principle. When one attacks a disordered
> state, then it will submit. And, if it submits, then the one who
> attacks will profit. When one attacks a disordered state, then he is
> in the right. And being righteous, the one who attacks will be
> glorified. If glory and profit are something the middling ruler
> would go in for, how much more so should this not be the case
> with a worthy ruler? Thus, the hacking up of one's land (i.e.,
> abdicating a piece of one's land to appease a potential attacker,),
> precious vessels, humble speech, and submissive behavior are
> insufficient to stop an attack. Only order is sufficient. If the state
> is ordered, then those who pursue profit, and those who pursue
> fame will not attack. (13.8a–9b and 13.10b)

The first section of this passage seems unequivocally to be saying
that the periodic sequence of the five elements continues through its
cycle regardless of man's reaction. The closing of this section clearly says
that the cycle passes through the final element and returns to the beginning, even, perhaps, if men do not recognize that final element./4/
Moreover, the beginning of the remaining section of the passage states
that the moving force behind the influence of this cycle on human
affairs is the affinity of things of similar natures for each other. So far,
the view of the influence of the sequence of natural phenomena on
human affairs seems to be analogous to a man floating down a river in a
boat. He may, if he is aware of the nature of the currents, steer accordingly and travel safely downstream. But he cannot go upstream, and if
he is unaware of the nature of the currents, he will run aground or meet
with some other difficulty. Yet, the passage then raises some doubt as to
whether or not an entirely mechanistic view of the influence of the cycle
of natural phenomena on human affairs is justified, for we are told,
"Heaven's sending down calamities and manifesting auspicious omens
both have a function." This remark surely implies a teleological view of
the role of the cycle of natural phenomena. That is, there may be a
purposeful "force" behind the ill consequences of not reacting properly
to the natural powers.

Adding to the ambiguity of this problem are the twelve *p'ien* which
begin each of the first twelve *chüan* 卷 of the *Lscc*. Each of these twelve
p'ien, all of which are also contained in the *Li-chi* 禮記 under the title,
"Yüeh-ling" 月令 , presents all the information regarding which natural
power is dominant in each month of the year, as well as which colors,
musical notes, etc., are connected with the various months. And each of
these *p'ien* closes with a statement of the disastrous consequences that
would result from improper responses to the dominant power, particularly with regard to improper sacrifices. Generally, the evil consequences
take the form of unseasonable weather. On the one hand, one could

argue that improper responses to the seasons, as, for example, planting spring crops in winter, would bring about mechanistically caused detrimental results. On the other hand, if the cycle of natural powers continually passes through its period regardless of man's activities, why sacrifice at all? This ambiguity is inherent in the *wu-hsing* cosmology. Fung Yu-lan noted this difficulty, saying, "Are the influences imparted by human affairs upon the seasons, mechanically produced and received? Or are they awesome manifestations produced by some God who becomes angry because the human sovereign acts in a way that is improper? The former view implied a mechanistic universe: the latter, a teleological one. The *Yin-yang* and Five Elements school seems never to have realized the breach lying between these two concepts, and has always vacillated between them, sometimes toward one, sometimes toward the other."/5/

In any case, my purpose in this section of the paper has been to demonstrate the *Lscc*'s espousal of a belief in an unaltering sequence of naturally occurring phenomena. Certainly, the "Ying-t'ung" *p'ien* quoted above does that. Moreover, the frequent references in this passage to the importance of the proper response to these natural powers implies the cycle does influence human affairs. Further, the statements regarding such unseasonable occurrences as snow in spring, which conclude each of the opening *p'ien* of the first twelve *chüan* of the text suggests man's behavior can, and does, influence the cycle of natural powers as well. Thus, it has been demonstrated that the *Lscc* does espouse a belief in an unaltering sequence of natural phenomena, and that that sequence and human actions exert influence on each other.

Demonstrating the *Lscc*'s view of the historical process as being one of a steady decline in the human condition presents some difficulty. There is no shortage of instances in the text where the past is declared to be superior to the present, or the present inferior to the past. For example, in a *p'ien* entitled "Huan-tao" 圜道 , it says,

> Yao and Shun were worthy rulers. Both of them took the worthy as their successors. They were unwilling to give the succession to their descendants. Yet, when they appointed an official they made sure he kept to his square. The rulers of the present age all wish the succession not to be lost, and they give it to their descendants. When they appoint an official, they are not able to keep him in his square. Through their personal desires they have made things disorderly. Why is this? What they desired rested in the remote future, while what they knew rested in the immediate present. (3.19a-b)

> The present generation is turbid to the extreme. The suffering of the people could not possibly be increased. The Son of Heaven has been cut-off. Worthies are neither seen nor employed. ("Chen-luan" 振亂 , 7.7b)

The law of the Former Kings says, 'Reward the one who does good. Punish the one who does evil.' This is the Way of antiquity. It cannot be changed. Now, what is right is not distinguished from what is not right. . . . There is no greater wrong than this. ("Chin-se" 禁塞 , 7.11a-b)

If rulers are worthy and the age is one of order, then those who are worthy are in the highest positions. If rulers are not worthy and the age is one of disorder, then those who are worthy are in the lowest positions. Now, the Chou house is about to be obliterated, and the Son of Heaven has already been cut-off. Of disorders, there are none greater than lacking a Son of Heaven. When there is no Son of Heaven, then the strong defeat the weak, and the many oppress the few. The military is used for mutual destruction, and it cannot be stopped. The present generation is this sort of time. ("Chin-t'ing" 謹聽 , 13.17a-b)

Clearly, the passages quoted above do provide support for the contention that the authors of the *Lscc* regard the past as superior to the present. Yet, there are in the text two *p'ien* which apparently contradict this view. Indeed, in their discussions of the *Lscc*, both Hu Shih 胡適 and Lo Ken-tse 羅根澤 have noted the apparent contradictions in these passages, and each has attempted to provide an explanation for them. Hu and Lo give different reasons for the apparently contradictory nature of these two passages, but they both clearly view the statements contained in these two *p'ien* as genuine lapses on the part of the authors of the text. I do not. However, before I summarize the arguments employed by Hu and Lo, and present my own explanation of the problem, I will present translations of the relevant sections of the two problematic passages. The "Ch'a-chin" 察今 *p'ien*, which contains a great deal of material of a seemingly contradictory nature, says,

Why should the ruler not imitate the methods of the Former Kings? It is not because they are not good, but because they are unobtainable for imitation. The methods of the Former Kings have come down through past generations, and men have added some to them, and taken some away from them. So, how could one obtain them and imitate them?

Even if men had not added to, or taken from them, still, they could not be obtained and imitated. The decrees of the eastern places, and the laws of past and present use different expressions and different principles. Thus, of the ancient decrees, many are not compatible with the expressions of the present. And, many of the laws of the present are not compatible with those of the past. People with different customs have a similarity to this. What they desire to do is the same, but what they do is different. . . . How, then, can one get a hold of and imitate the laws of the Former Kings? And, even if one could get a hold of them, it is still so that one should not imitate them, for they have a necessity for their time, and their time did not come down to the present with the laws. So, even if they came down to the present, still they could not be imitated.

Therefore, one tries to explain the Former King's making of their laws, and imitates what they used to make their laws. Just what was it that the Former Kings used to make their laws? It was Man. Oneself is also a man. Therefore, if one examines oneself, then one can understand others. If one examines the present, one can understand the past. The past and the present are one. Others and oneself are the same.

Those men who possess the *Tao* value using what is near to know what is far, using the present to know the past, and adding to what they see to know what they cannot see. (15.31b–32b)

After this passage follow some examples of using what is near, or small, to understand what is far, or big. Then comes an anecdote concerning an attempted invasion of the state of Sung by the state of Ch'u. Men were sent out by Ch'u to mark fords on a river the Ch'u army intended to cross at night. But, unbeknownst to the Ch'u army, the river flooded, thereby invalidating the markers. As a result, when the Ch'u army attempted to ford the river the consequences were disastrous. The text then continues,

When the rulers of the present imitate the laws of the Former Kings there is a similarity to this, since the times certainly are different from those of the Former King's laws.

So, when it is said, 'These are the laws of the Former Kings!' and one imitates them, how is governing on this basis not lamentable? Thus, if in governing a country there were no laws at all, then there would be disorder. Yet, if one preserved the laws and allowed no changes, that would be unreasonable. Neither unreasonableness, nor disorder can be used to maintain a country. As generations change and times change, changing the laws is fitting. It is analogous to a good doctor. As an illness goes through ten-thousand changes, so his drugs must make ten-thousand changes. If the illness changes, and the drug is not changed, a person who once had longevity, now is a child who died prematurely. Thus, in undertaking all affairs, it must be done in accordance with the law. But, the one who changes the law must, based on the circumstances of the time, make changes. If one uses this principle, there will be no errors in affairs.

The common people are the ones who dare not even discuss the laws. The ones who must protect the laws unto death are the officials. Those who change the laws according to the times are the worthy rulers. This is the reason that there have been seventy-one sages in the world and their laws were all different. It was not intentional that they differed from each other, but their times and circumstances were different. (15.33a–34a)

The relevant passage from the second *p'ien*, which is entitled "Ch'ang-chien" 長見 , is much shorter than the piece presented above. It says only,

> The reason one person's wisdom overcomes another's wisdom is based on their farsightedness compared to their shortsightedness. The present's relationship to the past is like the past's to its future. And the present's relationship to its future is also like the present's to its past.
>
> Therefore, carefully learn to know the present and one can know the past. And if one knows the past, then he can know the future. Past and present, before and after are one. Thus, the sage understands a thousand years into the past, and a thousand years into the future. (11.13b–14b)

As mentioned above (p. 112), Hu Shih and Lo Ken-tse both feel the two pieces just presented contradict the view espoused so frequently in the text that the past is superior to the present. Undeniably, both passages explicitly state that in some respects the past and present are the same. Moreover, as Hu Shih points out in his discussion of these two pieces,/6/ the "Ch'a-chin" *p'ien* seems to contradict itself by declaring at one point that the circumstances of the past have not come down to the present, and at another point that the past and present are one. Hu argues that both these problems stem from the fact the authors of the *Lscc* were influenced by the philosophies of Hsün-tzu 荀子 and Han Fei-tzu 韓非子 . According to Hu, Hsün-tzu's argument for not using the Former Kings as a model was essentially that given at the very beginning of the "Ch'a-chin" *p'ien* presented above—namely, that the extant records of the actions taken by the Former Kings were made deficient during transmission from the past to the present. Further, Hu says Hsün-tzu had no concept of historical "evolution" (*li-shih yen-hua* 歷史演化)./7/ For Hsün-tzu, past and present are the same, but only the recent past, that of Hsün-tzu's so-called "Later Kings" 後王 , was adequately preserved in historical records to be used as a model. Thus, according to Hu, where the authors of the *Lscc* speak of the past and present as one, they are reflecting the influence of Hsün-tzu.

On the other hand, the view that the circumstances of the past have not come down to the present, and hence, the methods of the past are inapplicable to the present, the espousal of which constitutes the majority of the text of the "Ch'a-chin" *P'ien*, is a manifestation of the influence of Han Fei-tzu's philosophy on the authors of the *Lscc*. This is undoubtedly true. The similarities between the arguments found in the "Wu-tu" 五蠹 *p'ien* of the *Han Fei-tzu* and the "Ch'a-chin" *p'ien* of the *Lscc*. are remarkable. In concluding his explanation of the apparent contradictions in these two sections of the *Lscc*, Hu makes it clear that he feels the contradictions caused by borrowing from Hsün-tzu and Han Fei-tzu were an oversight on the part of the authors by saying that when the text declares the circumstances of the past have not come down to the present, and hence, the past cannot be used as a model, that is, "the real significance of the book's advocating not imitating the Former Kings."/8/

Lo Ken-tse, in his article entitled, "An Examination of the Rejection of the Past Among Late Chou Philosophers" 晚周諸子反古考 , which is contained in the sixth volume of the *Ku-shih pien* 古史辨 /9/ argues differently. Lo's article is actually concerned with a refutation of K'ang Yu-wei's 康有為 thesis that the late Chou philosophers all relied on the past. In the final section of this article, which deals with the *Lscc*,/10/ Lo defines the the two terms "T'o-ku" 託古 and "fan-ku" 反古 as referring to those who view the past as superior to the present, and those who view the present as superior to the past respectively./11/ Further, he declares that both the "Ch'a-chin" and "Ch'ang-chien" sections of the *Lscc* clearly espouse the *fan-ku* position. That is, according to Lo, the "Ch'a-chin" and "Ch'ang-chien" sections both maintain a view of the present as being superior to the past. This specific assertion is simply not true. Even the examples cited by Lo himself do not support his view. What is said in both these sections is that the past and the present are, in some respects at least, the same. Nonetheless, the question remains, do the statements in these two sections concerning the relationship between the past and the present really constitute a contradiction of the basic tenet of the text that the present represents a decline in the historical process?

The answer to this question is no. What has happened here is that the level of the discussion in the text has changed. Whenever one maintains the position that change is constantly occurring, as the authors of the *Lscc* do, that position itself entails a paradoxical sort of contradiction if care is not taken to specify that two levels of discussion are present. For example, if one says, "Change is constant," that can be taken to mean either of two things. If the statement is viewed as operating on a single level, it could be taken to mean, "Change does not change." On the other hand, if one is aware that the statement is concerned with discussing the nature of principles which underlie change, then the statement could more adequately be taken to mean, "Change is what constantly occurs." When the *Lscc* authors say, as they do in both the "Ch'a-chin" and "Ch'ang-chien" sections of the text, that the past and present are one, I believe they are simply attempting to demonstrate they they are aware of the constant nature of change. Indeed, it is surely justifiable to take the discussion of the relationship between past, present, and future to be an attempt to show that all changes must be somewhat relative. That is, they are declaring that they are aware that there are no "eternal answers" for alleviating the ills of the human condition because the present is the future of the past, and the past of the future. Therefore, just as they argue that the "cures" of the Former Kings are not applicable to the present, so they would argue that any "cures" which might be used in the present, would not work in the future.

Thus, in those places where the text makes clear its belief that the present is inferior to the past, the level of discussion is the actually existing physical world, wherein the principles underlying change manifest themselves; and the inferiority of the present to the past is the actual result of that manifestation. It is not a characteristic of the view of historical change put forth in the *Lscc* that the passage of time must bring about a decline in the human condition—that the past must always be superior to the present. On the contrary, the intention of the authors is only to show that up to the time of the compilation of their text, this has been the case, but if their advice is accepted, then the future could well be better than the past.

On this point, any discussion of the nature of change is likely to seem contradictory. When one says x causes y change, two notions are present. Y is a constantly changing phenomenon, and x always IS. The x of the *Lscc* is the constant succession of the Five Elements through the unalterable sequence of their cycle. As was shown in the earlier sections of this paper, which dealt with the influence of the human actions on the effects of the cycle of natural phenomena, the factor which keeps y itself from becoming simply a mirror image of x is human response to that cycle.

This problem in treating change is certainly not unique to Chinese philosophical speculations. In Plato's dialogue, the *Phaedo*,/12/ four arguments are given for the immortality of the soul. The first of these, the so-called "cyclical argument" (69E–72E) hinges on the assumption that opposites generate opposites. Therefore, according to the argument the living generate the dead, and the dead the living. Then, toward the end of the dialogue, Plato has Socrates declare all the previous arguments inadequate (95E–99D) on the grounds that they operate on the physical level. He then introduces a "new method" for dealing with the expressly nonphysical nature of this higher level (99D–102A). Then he proceeds with his final proof for the soul's immortality. At the outset, he points out that a physically existing entity might partake of opposite qualities simultaneously. For example, a man might be taller than one man and shorter than another, thereby "partaking" of "tallness" and "shortness" at the same time. However, "tallness" itself cannot partake of "shortness," nor "shortness" itself, "tallness." At this point (103A), someone in the audience reminds Socrates of his opening "cyclical argument," wherein he argued that opposites are generated from their opposites. Socrates replies,

> But don't you realize the difference between what we are talking
> about now and what we were talking about then? *Then* it was

said that an opposite thing is generated from an opposite thing, but *now* that the opposite *itself* could never be opposite to itself—either that which exists in us or that which exists in nature. Then we were talking about things which *possess* the opposites, calling them by the same name as the opposites themselves have, but now we are talking about those opposites themselves, which by their presence, give their names to the things called after them; and we say that they themselves would never submit to becoming one another. (103B)

Clearly Plato, in the "cyclical argument," was speaking of an entity, a human being, "partaking" of life and death. Then, when he moved from the physical level to the metaphysical, he had to emphasize that life "itself" simply could not partake of death "itself" and vice versa. So too, in the *Lscc*, the authors had to emphasize that it was change which was constant, not the apparent decline in the human condition that they saw as the manifestation of change up to their times.

Moreover, the main intent of the text was not to discuss historical change, but to discuss how to deal with its effects. In this discussion the consistency of the ideas being put forth becomes clear. I have shown that the authors of the *Lscc* viewed change as a product of the interaction of the cycle of the Five Elements and concurrent human actions. Moreover, it was clearly demonstrated in the section which dealt with the nature of this interaction that the authors of the *Lscc* believed heaven manifested signs to make known which element was dominant. Thus, since the authors were equally explicit in stating the sequence of the cycle, one can easily deduce which element will succeed the one which is presently dominant. Therefore, if one is observant one can always know how the cycle of elements is functioning at any given moment.

Further, as both the "Ch'a-chin" and "Ch'ang-chien" sections translated above declare, there are certain elements basic to human nature which are the same in all men. Therefore, if one knows himself, he can know others. The point is reiterated many times, and one *p'ien*, entitled "Tzu-chih" 自知 (24/6a–8a), takes this thesis as its topic. Now, if historical change is taken as a product of the interaction of natural phenomena, which are regulated by a knowable cycle, and of the actions of men, who are impelled by knowable elements in their nature, then a man of great perception and comprehensive knowledge is capable of predicting changes. Indeed, we are told this very thing several times in the text. Both the "Ch'a-chin" and "Ch'ang-chien" passages presented above state this. And there are additional examples in the text. In a *p'ien* entitled, "Ch'a-wei" 察微 (16.20a–25a), we are told how careful observation of the significance of seemingly minor occurrences enabled perceptive observers to foretell major consequences. But, the clearest example of all is contained in a *p'ien* appropriately entitled, "Chih-hua"

知化 (23.7a–9a). The entire *p'ien* is devoted to emphasizing one point—change not only can be known, it must be known. If one does not know what sort of change is going to occur before it happens, he may as well not bother to know it at all.

Clearly, knowledge which includes knowing the past, present, future, oneself, and hence others, and the influence the cycle of Five Elements is having, and will have, is knowledge of the most comprehensive sort. At one point, (16.22a), the "Ch'a-wei" *p'ien* says, "In maintaining a state, the very best is to know the future, next is to know the past, and next is to know the present." Examples which place more emphasis on the necessity of comprehensiveness in one's knowledge can be found in: the "Huan-tao" *p'ien*, where it says "Heaven's *tao* is circular. Earth's *tao* is square. The sage-ruler imitates them" (3.17b); the "Hsü-i" 序意 *p'ien*, where the message is that the ruler must imitate heaven and earth (12.14a-b); and the "Yu-shih lan" 有始覽 which says, "Thus all things are made complete. Heaven makes all things flourish. The Sage observes them so as thereby to examine his own kind. He finds the explanation of how Heaven and Earth became concrete forms, how thunder and lightning are produced, how *yin* and *yang* form the essence of things, and how people, birds and beasts are in a state of peace" (13.6b–7a)./13/ The passages cited on the preceding pages should be adequate to serve as evidence for the contention that the authors of the *Lscc* do view change as knowable. Further, implicit in many of the passages cited is the contention that the only means available for gaining a knowledge, or perhaps I should say a foreknowledge, of change is comprehensive understanding.

To summarize the manner in which I regard the notions of "historical change" and "comprehensiveness" as functioning in the *Lscc*, the characteristic I would emphasize is that the two notions are integrated into a consistent, systematic view of historical change and how a ruler ought to deal with it. Historical change is knowable. The means by which one can come to know the nature of historical change is comprehensive knowledge. With his knowledge of historical change, made possible through his possession of comprehensive knowledge, the ruler will recognize the influences of all the factors affecting the course of historical change. Hence, one can easily predict the future. Thus, proper responses to each circumstance that arises are possible, and no calamity can befall him or his state.

NOTES

/1/   All references to the *Lü-shin ch'un-ch'iu* are to: Hsü Wei-yü 許維遹 , ed., *Lü-shih ch'un-ch'iu chi-shih* 集釋 (Peking, 1935), hereafter abbreviated as *Lscc*.

/2/ Fung Yu-lan, *A History of Chinese Philosophy*, trans. by Derk Bodde (Princeton, 1953), 1:161–62. Cf. Wing-Tsit Chan, *A Source Book in Chinese Philosophy* (Princeton, 1963), p. 250, and Joseph Needham with Wang Ling, *History of Scientific Thought*, vol. 2 of *Science and Civilization in China* (repr. Taipei, 1971), p. 238. It should be noted that both Fung and Needham clearly feel that the authors of the *Lscc* are indebted to the *wu-hsing* philosophy of Tsou Yen 騶衍 for this passage. also noteworthy is the last line which could be taken as a prophecy of the fall of the Ch'in. Needham renders this and the ensuing line as: "And that dispensation will inturn come to an end, and at the appointed time, all will return once again to Earth. But when that time will be we do not know."

/3/ All four of these sentences rendered using "are like" are actually four character phrases lacking a verb. Justification for this rendering comes from *Huai-nan tzu*, *Sppy*, 6.2a. Further, the discussion of notes above and the military below appears in *Lscc*, 20, 12b–13b.

/4/ There is some question as to how the phrase *pu-chih shu-pei* 不知數備 (13.7b) ought to be translated. Both Needham (note 2 above) and Wilhelm (Richard Wilhelm, trans., *Frühling und Herbst des Lü Pu-we* [Jena, 1928], p. 161) render it differently, though I take the sense of Wilhelm's translation to be essentially the same as Bodde's.

/5/ Fung, *History of Chinese Philosophy*, vol. 1, p. 165.

/6/ Hu Shih, "*Tu Lü-shih ch'un-ch'iu*" 讀呂氏春秋, in *Hu Shih wen-ts'un san-chi* 胡適文存三集 (Shanghai, 1930), p. 388.

/7/ Ibid.

/8/ Ibid.

/9/ Lo Ken-tse, "Wan-Chou, chu-tzu fan-ku k'ao," in *Ku-shih pien*, ed. by Lo Ken-tse (Shanghai, 1938), 6:1–49.

/10/ Ibid., pp. 45–49.

/11/ Ibid., p. 47.

/12/ The line numbers given throughout this discussion of the *Phaedo* are standard manuscript numeration. Two editions were used as references: R. S. Bluck, trans., *Plato's Phaedo* (London, 1955) and W. D. Geddes, ed., *The Phaedo of Plato* (London, 1885; 2nd ed.).

/13/ This passage was translated by Derk Bodde in Fung, *A History of Chinese Philosophy*, vol. 1, p. 168.

Concepts of Comprehensiveness and Historical Change in the *Huai-nan-tzu*

Jeffrey A. Howard

Introduction°

The *Huai-nan-tzu* 淮南子 (hereafter *HNT*) was the last work of classical philosophical Taoism; and it forms a part of the beginning of a new genre of Chinese philosophical and historical writing which included the *Lü-shih ch'un-ch'iu* 呂氏春秋 and ultimately the *Shih-chi* 史記 . This new genre attempted to be comprehensive in its coverage of material with certain underlying themes which the material served to reinforce. Standing as it did at the end of one tradition and the beginning of another, the *HNT* is an invaluable source for this period of change in the history of Chinese philosophy.

The authors of the *HNT* had a very specific program and they particularly utilized this new view of comprehensiveness in addition to a dynamic view of historical change. The rather static worldview of Hsün-tzu 荀子 was replaced by one which incorporated the relativity of Chuang-tzu 莊子 with the concept of changing with the times put forth most forcefully by Han Fei-tzu 韓非子. The authors quite openly discuss the techniques that they use in their attempts to win over their readers, the most important of which are: comprehensiveness, investigation, and historical change. All of these areas are interrelated within the scheme of the *HNT*. Comprehensiveness is necessary for an appreciation of the unity of the *Tao* but is impossible without investigation. Investigation must be carried on all around, and especially in the past, from which *HNT* has drawn historical examples and models to illustrate various precepts of the functioning *Tao*. Investigation and comprehension of the past leads to a knowledge of the separateness of time; time is an entity distinct from man and for actions to be proper they must be in accord with the proper time. Only comprehensiveness gives the necessary perspective./1/ This scheme was used in an appeal to the reader, challenging him to realize that everything, all classifications of things, partake of the *Tao*.

In the following paper I will examine the concepts of comprehensiveness, investigation, and historical change as used in the *HNT* in its efforts to explain the manifestations of the *Tao* to readers who apparently demanded more than the metaphysical explanations of the *Tao-te Ching* 道德經 and the *Chuang-tzu*. In a paper of this size, I have dealt basically with the *HNT*/2/ itself and have touched only briefly on its relations to other texts or to the period during which it was written.

Background

"THEY KNOW ONE THING,
BUT THEY ONLY KNOW THAT ONE."
(*Shih Ching*, Mao 195.6. Legge, IV, p. 333)

These lines from the *Shih ching* are quoted in the *HNT* (*HNHL* 8/9b) as a criticism against those who have less than comprehensive knowledge. This criticism reflects the theme of the entire work; through comprehensive knowledge, the unity of all things in the *Tao* may be known. The format of the work, twenty essays, at times approaching monographs on various subjects (e.g., #3 "T'ien-wen" 天文 "Patterns of Heaven," #4 "Ti-hsing" 墬形 "Forms of the Earth") with a twenty-first chapter which ties the entire work together, is well-suited to the work's stated purpose: "In setting down these writings and discussions I (we) have sought to illuminate and encompass the way and its inner power, and to make clear the web of human affairs . . . I (we) have covered the essentials and presented a general picture . . . therefore I (we) have written these twenty chapters" (HNHL 21/1a. *Sources*, p. 185). The work was composed by scholars at the court of Liu An 劉安 (180?–122 B.C.), many of whom possibly found the atmosphere surrounding Emperor Wu 武 (r. 140–86 B.C.) intolerable, and for that reason fled to the court of a patron more appreciative of their merits./3/ There has even been a suggestion that the eventual fate of Liu An was due in part to jealousy on the part of Emperor Wu toward Liu An's intellectually stimulating court./4/

These essays themselves were reputed to be the work of eight scholars/5/ and the contribution of Liu An is impossible to judge, although it is certain that he was an accomplished man of letters./6/ There is no question that Liu had Taoist leanings and even aspired to a certain immortality as seen in the *Han shu* 漢書 biography of Liu Hsiang 劉向 (77–6 B.C.) who was entrusted with materials traced to Liu An which were in part alchemical recipes for longevity./7/ However, immortality practices play no significant role in the *HNT*; much of the Taoism apparent in the work is tempered somewhat from the idealistic work of Chuang-tzu and the problematic text of Lao-tzu, and is placed within the context of the real world.

The work was somewhat overshadowed by that of Tung Chung-shu 董仲舒 (179?–104? B.C.), and it was not until the Later Han (A.D. 25–220) that attention was drawn to it. In the Later Han period, Ma Jung 馬融 (A.D. 79–166), and Hsü Shen 許慎 (A.D. 30–124), wrote commentaries on the work, but that of Ma Jung is now lost. That of Hsü Shen and of Kao Yu 高誘 (fl. A.D. 212) now exist, but in a form which is somewhat confused./8/

Comprehensiveness

Comprehensiveness is both the goal and the method of the *HNT*. While it is a Taoist unity of things that ties the work together, the unity is perceived only through a complete investigation and comprehension of all things. By such an investigation and comprehension the transformations of things can be known and the actions of an individual, either as an individual or in the capacity of ruler, will be based on the unity of the *Tao*, the source of all change. That the work is directed at the cultivation of both the outward actions of man (affairs) and the inward man himself is evident from the postface of the *HNT* (ch. 21): "In general what is collected in this book is the means to observe the *Tao* (in its) openings and closings, with a view toward enabling the later generations to understand (what is) fitting in upgrading and degrading, selecting and casting aside. Externally it is a means to come into contact with things without becoming confused (in them). Internally it has the means to settle the spirit and nourish the life breath" (*HNHL* 21/6a). This in essence is the basis of the entire collection of essays in the *HNT*. How does this conception of comprehension and investigation compare with earlier conceptions of the same within the Taoist philosophical tradition?

The *Tao-te ching* deprecates the investigation of things:

> Without stirring abroad
> One can know the whole world;
> Without looking out of the window
> One can see the way of heaven.
> The further one goes
> The less one knows.
> Therefore the sage knows without having to stir,
> Identifies without having to see,
> Accomplishes without having to act./9/

Here, the deeper the evaluating mind/10/ becomes involved in the multiplicity of things, the further it moves away from the unity of the *Tao*.

The *Chuang-tzu* considers that this unity of the *Tao* is the thing to which the mind must hold; it is senseless "to wear out your brain trying to make things into one without realizing that they are all the same."/11/ The unevaluating mind is the way to perceive the unity:

> Whether you point to a little stalk or a great pillar, a leper or the beautiful Hsi-shih, things ribald and shady or things grotesque and strange, the Way makes them all into one. Their dividedness is their completeness; their completeness is their impairment. No thing is either complete or impaired, but all are made into one again. Only the man of far-reaching vision knows how to make them into one. So he has no use for categories, but relegates all to the constant. The constant is the useful; the useful is the passable; the passable is the successful; and with success, all is accomplished. He relies upon this alone, relies upon it and does not know he is doing so. This is called the Way./12/

The goal of the *HNT* is similar—realization of the unity in the *Tao*—but the method of attainment is quite different. The program set forth in the postface to the *HNT* depends much less on abstract presentation, such as is found in the *Tao-te ching* and the *Chuang-tzu*, and much more on investigation of the real world in a comprehensive manner. At the very beginning of the postface, the author(s) of *HNT* say: "Thus if I (we) spoke only of the Way and did not mention human affairs, one would not be able to get along in the world today, while if I (we) spoke too much of human affairs and did not discuss the Way, one would not be able to move and rest with the transforming process" (*HNHL* 21/1a; *Sources*, p. 185).

In the *HNT* the discussion of "human affairs" is the means and the "Way" the end. The Way is essentially that of the unity of the *Tao* found in the *Tao-te ching* and the *Chuang-tzu*, but the means, in many cases, are much more firmly rooted in the practical illustrations drawn from human affairs. This is not to say that each chapter bases itself on a firm pragmatic approach with examples drawn from historical sources. While this is the case with a number of chapters, there are also those that are similar to the *Chuang-tzu*. But the author(s) are aware of this problem and in the opening passages of the postface say: "I have gazed upward to study heaven, examined the earth below me, and about me sought understanding of the principles of humanity. Though I have not been able to draw forth the heart of the supreme mystery, from the abundance I have presented, one may trace its course and outline" (*HNHL* 21/1a; *Sources*, p. 185). The work seeks to be comprehensive in both method and scope and seeks to impart comprehensiveness. In "gazing upward to study heaven" the work covers astronomical phenomena, seasons, weather, etc., and relates them to the circumstances on the earth. In "examining the earth below me" the work covers both the geographical configurations of the earth in addition to the interworkings of the five elements, and in "about me seeking the understanding of the principles of humanity," the book draws on the world of men to provide examples of proper and improper understanding of the great principles of unity. "Therefore my words have been many and my explanations broad. Again I fear that men may depart from the root and seek after the branches" (*HNHL* 21/1a; *Sources*, p. 185). The *HNT*, then, seeks to be comprehensive in order that men will have every opportunity to understand the philosophy that is presented.

Investigation

In the *HNT*, comprehensiveness, which will lead to an appreciation of the unity of the *Tao*, is arrived at not by the unevaluating mind of the *Tao-te ching* and the *Chuang-tzu*, but by the evaluating, investigative mind: "One who investigates (only) one affair and masters only one talent is a middling man. One who covers comprehensively and embraces totally, who estimates ability and utilizes it, is a sage" (*HNHL* 10/18a); or further: "One cannot speak about transformation with one who has investigated only a single detail. One cannot speak about greatness with one who has examined only one event" (*HNHL* 10/17a). This passage is reminiscent of the frog in the well discussing the sea with the great turtle./13/ Relativity, which in both the *Chuang-tzu* and *HNT* eventually leads to an understanding of the unity of the *Tao*, can be obtained through experience; and in the *HNT*, this means an investigation into the multiplicity of things. The entirety of the eleventh chapter of *HNT*, "Ch'i-su" 齊俗 "Unifying Customs," attempts to point out this relativity through a discussion of the variations of man's customs. The postface of *HNT* describes this eleventh chapter as follows: "The chapter 'Ch'i-su' unifies mankind's defects and excellencies, harmonizes the customs of the nine tribes, consolidates the discussions of past and present and connects (together) the principles of all things" (*HNHL* 21/3a).

Investigation is also considered an important virtue for the ruler; in the ninth chapter, "Chu-shu" 主術 (Craft of the Ruler), we find: "Now the traces of the originating power of order and disorder can be seen but there are no rulers in the world who investigate them. This is what blocks the *Tao* of order" (*HNHL* 9/14b). Nothing, according to the *HNT*, should elude the comprehension of the ruler: "The division of (something as small as) one inch can be investigated if light is borrowed from a mirror in order to illuminate it. For this reason the eyes and ears of the enlightened ruler are not troubled; his spirit is not exhausted. As things arise he perceives their changes; as events are manifested he responds to their transformations. What is near is not in disorder and the distant is well governed" (*HNHL* 9/21a). It is through comprehensive knowledge that the ruler can choose worthy ministers who in turn are expected to investigate the principles behind affairs about which they speak (*HNHL* 9/13b). This is the basis for the ruler's acting in accordance with *wu-wei* 無為 . In the *HNT* *wu-wei* is a dynamic principle; the author(s) in the nineteenth chapter, "Hsiu-wu" 脩務 (Necessity of Training), state that they have never heard a sage describe *wu-wei* as serene, meditative nonaction (*HNHL* 19/1a; Morgan, p. 220). Rather, *wu-wei* is described as action which is free from selfish ideas or personal will or desire (*HNHL* 19/4b; Morgan, p. 224). In the sense that the ruler lets his ministers carry out his actions, he is acting in accordance with *wu-wei*; but for the ruler to choose his ministers, a comprehensive knowledge is required. As the *HNT* states in chapter 10, "Miu-ch'eng" 繆稱 (Erroneous Designations), those who are not themselves capable, cannot recognize capability in others (*HNHL* 10/3a).

Investigation, then, is the key to the attainment of comprehension and to comprehensiveness. Lack of investigation leads not only to loss of individual awareness but, in the case of government, to an inability to govern: "The governing methods of Shen Pu-hai 申不害 , Han Fei-tzu and Shang Yang 商鞅 uprooted the foundations, laid waste to the root and did not exhaustively investigate into the origins of life . . . in their austerity they turned their backs on the root of *Tao* and *Te*; in contending over trifles they killed off the people . . . but considered themselves to be governing" (*HNHL* 6/15b). This type of government of lessened virtue and increased punishment obviously would not attract people of worth and is likened by the *HNT* to holding up a slingshot to attract birds (*HNHL* 9/2b).

The investigation of the past is another aspect of the concept of comprehensiveness in the *HNT*. Liu Chih-chi 劉知幾 (A.D. 661–721) was aware of the historical scope of the work when he wrote in his *Shih t'ung* 史通 : "Formerly, during the Han, Liu An wrote a book called the *Huai-nan-tzu*. The book pens up the whole world, encompassing the extremities of the ancient and modern (periods) all the way from T'ai Kung (ca. 1122 B.C.) down to Shang Yang (d. 338 B.C.)."/14/ In its attempts to explain difficult concepts like *Tao* and *Te* and the unity of all things, the *HNT* uses a wide range of historical examples as models of understanding. For example, the discussions in the second chapter, "Ch'u-chen" 俶真 (Beginning the truth), range in time from Fu Hsi 伏義 through Shen Nung 神農 and Huang Ti 黃帝 down to the decay of the Chou 周 house, and culminating in the rise of the philosophies of Confucius, Mo-tzu 墨子 , and Yang Chu 楊朱 . In the sixth chapter, "Lan-ming" 覽冥 (Peering into the Mysterious), examples of historical affairs are drawn from a period just prior to the composition of the *HNT*. In one case the chapter draws on a story concerning Meng Ch'ang-chün 孟嘗君 (d. 279 B.C.) who as minister in Ch'i led campaigns against the state of Ch'in 秦 (*HNHL* 6/2b). The eleventh chapter,

"Ch'i-su" (Unifying Customs) draws on historical personages such as the Duke of Chou 周公 (d. 1105 B.C.; *HNHL* 11/3a; Wallacker p. 30), Duke Huan of Ch'i 桓公 (r. 683–641 B.C.; *HNHL* 11/7b; Wallacker p. 34), Duke Hsien of Chin 獻公 (r. 675–650 B.C.; *HNHL* 11/8b; Wallacker p. 35), King Chuang of Ch'u 莊主 (r. 612–589 B.C.; *HNHL* 11/9b; Wallacker p. 35) and others. The twelfth chapter, "Tao-ying" 道 應 (The Way and its Effects), examines certain passages from the *Tao-te ching* in light of historical incidents which illustrate each particular passage. Here the method is concerned with gaining insight into the philosophy of Lao-tzu by "grasping up the traces of affairs that have had their course" (*HNHL* 21/4a), that is, through historical illustration. Of the fifty-four separate illustrations/15/ the majority derive from historical incidents of the *Ch'un-ch'iu* 春秋 period (722–418 B.C.). The thirteenth chapter, "Fan-lun" 氾論 (General Discussions), lists a number of propositions which were held at certain times to be either proper or improper, but which, with the changing of time, were themselves reversed. The propositions are illustrated by a number of historical incidents, among them the surprise attack on Cheng by Duke Mu 穆公 of Ch'in, the rise of Su Ch'in 蘇秦 , Kuan-tzu 管子 becoming leading minister of Ch'i and others./16/ The fifteenth chapter, "Ping-lüeh" 兵 略 (Military Strategy), utilizes historical illustrations ranging in time from Huang ti (said to have ruled 2648–2548 B.C., *HNHL* 15/1b; Morgan, p. 183) to Erh-shih huang-ti 二世皇帝 (r. 210–206 B.C., *HNHL* 15/8a; Morgan, p. 193). In the eighteenth chapter, "Jen-chien" 人間 (Human Affairs). the illustrations cover a wide range of Ch'un-ch'iu and Chan-kuo historical personalities, for example: Duke Huan of Ch'i (*HNHL* 18/8a), Duke Chuang of Ch'u (*HNHL* 18/2b), Duke Li of Chin 厲公 (r. 225–210 B.C., *HNHL* 18/8a) and Li Ssu 李斯 (d. 208 B.C., *HNHL* 18/8a).

Throughout the work, the reader is exhorted to look into the past to discover the processes at work throughout history: "To expect a rich reward for an insignificant action or to accumulate ill will but not (expect) calamity is unheard of in the world. Therefore, if the sage investigates the results of the past, then he will understand what is to happen in the future" (*HNHL* 10/2a). The use of historical models is a very conscious device used by the authors of *HNT* to drive home their philosophical message. This is especially clear in the postface which says:

If one speaks of the *Tao* but does not understand the end and beginning then one does not understand what is to be imitated. If one speaks of the end and the beginning but does not make clear heaven and earth and the four seasons then one does not understand what is to be avoided. If he speaks of heaven and earth and the four seasons but fails to draw comparisons (from others) to take hold of things, then he does not understand the subtle refinement. (*HNHL* 21/6a)

The key words in this passage are "imitate," "avoid," and "draw comparisons." They are all the outcome of a close study of the models and illustrations which are provided by the *HNT*. Further, the postface says: "If we spoke of worldy changes but did not speak of past affairs, then one would not understand the (proper) reply of *Tao* and *Te*" (*HNHL* 21/6a).

Historical Change

But what of historical change? Did the *HNT* perceive of historical change as a dynamic process? Certainly, the *HNT* did not hold with the position as outlined in Hsün-tzu that "the ancient times and the present are alike. If the classification is not violated, although it is old, the principle remains the same."/17/ Given the sense of relativity gained by comprehensive knowledge of the multiplicity of things, the *HNT* recognized that things did change over time: "The virtue of the Three Dynasties was the accumulation of 1000 years of honor. The evils of Chieh and Chou were the accumulation of 1000 years of destruction" (*HNHL* 10/16b). Or again: "The loss of this generation's soul was a result of gradual decay; the causes were old" (*HNHL* 2/15b; Morgan, p. 48). There is even a hint of biological and geological change; in the seventh chapter, "Ching-shen" 精神 (Spirit and Soul), the development of an embryo is described from the second through the tenth month (*HNHL* 7/2a), and in the fourth chapter, "Ti-hsing" (Forms of the Earth), there is a treatment of the growth of metals in the earth (*HNHL* 4/17a–18b)./18/ However, the predominant sense of historical change is one of decline; in the beginning there was the unity of the *Tao*, but the golden age is lost and the decline has continued until the present age (*HNHL* 2/13–14b, 8/2a, 10/1b, 11/1a, 13/4b). The *HNT* describes in detail the earliest times in the various creation stories./19/ In these creation stories, the undifferentiated *Tao* was at the very beginning; a unity without division into classes (*HNHL* 2/13b, 3/1a), and even before time (*HNHL* 2/1a). In distinction to Hsün-tzu, who hoped to recapture the civilization of the former kings, to reinstitute the golden age in detail, the *HNT* hoped only to utilize the knowledge of undifferentiated unity which characterized those earliest times. There are no specific programs from the past which the *HNT* hopes to reinstitute. In fact, the concept of time as presented in *HNT* intimates that, because it is ever changing, there is need for different measures: "Time does not stop in the midst of its fluctuations; if too early, then one goes past it, if too late, then one cannot catch it" (*HNHL* 1/16a; Morgan, p. 15). Here, time is distinct from the world of man; a continuum which man must be aware of if he is to catch the correct moment in which to act. That awareness necessitates the kind of perspective that the *HNT* calls comprehensive.

Decline is the historical perspective in the *HNT*, and this sense of decline appears to have been the prime motivating factor in writing the *HNT*. As a counter to this decline, the *HNT* attempted to be comprehensive and to encourage comprehensiveness by exhorting its readers to investigate the world and to look into the past; to learn from historical models and to cultivate both the inner and outer person. The urgency with which the postface discusses the methods used to aid the reader appear to stem from the belief that the knowledge contained in the *HNT* was not available elsewhere:

> Books like this of Mr. Liu (Prince of Huai-nan) observe the forms of heaven and earth, penetrate the affairs of yesterday and today, weigh the curcumstances and set up their systems, measure the situation and set forth what is proper. Seeking the heart of the Way, according with the customs of the Three Kings, it brings together from far and wide. Into the center of the subtlest mystery it skillfully bores to observe its smallest part. Casting away the foul and impure, drawing forth the pure and silent, it unifies the world, brings order to all things, responds to change and comprehends classes and categories. It does not pursue a single path, nor guard one corner of thought, bound and led by material things and unable to change with the times. Use it in a corner of life and it will never fail; spread it over the whole world and it will never be found lacking. (*HNHL* 21/9b–10a; *Sources*, p. 189)

Epilogue

A *Shih ching* quotation, used critically by *HNT*, was cited at the beginning of the paper: "They know one thing, But they only know that one" (Mao 195.6: Legge IV, p. 333). The quotation is reminiscent of a fragment from the 7th c. B.C. Greek poet Archilochus: "The fox knows many things, but the hedgehog knows one big thing" (Diehl, *Frag* 103). Isaiah Berlin has used the hedgehog and the fox to represent certain polarities of thought; the hedgehog represents centripetal thinking which: "relates everything to a single central vision, one system less or more coherent or articulate . . . a single, universal, organizing principle in terms of which alone all that they are and say has significance."/20/ The fox is centrifugal and pursues: "many ends, often unrelated and even contradictory, connected, if at all, only in some de facto way . . . their thought is scattered or diffused, moving on many levels, seizing upon the essence of a vast variety of experiences and objects for what they are in themselves."/21/ Berlin's use of the fox and hedgehog, or I should say his interpretation of Archilochus's fragment, is an interesting way in which to examine polarities of Western thought; however, this kind of examination of thought is not unknown to Chinese speculation and even helps to describe the *HNT*.

Chang Hsüeh-ch'eng 章學誠 (A.D. 1738–1801) indulges in much the same kind of analogy as that of Berlin when he considers that Ssu-ma Ch'ien's 司馬遷 (145–90? B.C.) *Shih-chi* was like a circle/22/ and partook of attributes which Nivison has suggested were flexibility, insight, and spirit./23/ Chang considered Pan Ku's 班固 (32–93 A.D.) *Han shu* 漢書 to be square/24/ and essentially more concerned with record keeping, exactness, and wisdom,/25/ If I may be allowed to indulge in a bit of cross-cultural speculation, I would consider Pan Ku to be a hedgehog/26/ and Ssu-ma Ch'ien a fox.

To carry the analogy further, it is this fox quality which characterizes *HNT* and the other texts mentioned at the beginning of this paper. What characterizes Berlin's use of Archilochus's fox is his comprehensive knowledge, precisely the kind of thinking that we see in such works as the *Lü-shih ch'un-ch'iu*, the *HNT* and Ssu-ma Ch'ien's *Shih-chi*. This same quality is found in later historical critics who praised Ssu-ma Ch'ien and disparaged Pan Ku, notably Cheng Ch'iao 鄭樵 (A.D. 1104–1162) and Chang Hsüeh-ch'eng. The *HNT* represents one of the earliest examples of what was to be a recurring theme in Chinese historical criticism; the need for a comprehensive overview, which would display developments over time.

NOTES

°On texts and translations: I have used the *Huai-nan hung-lieh chi-chieh* 淮南鴻烈集解 of Liu Wen-tien 劉文典 (Shanghai, 1933), hereafter cited as *HNHL*, as the basic text of *HNT*. Where the chapter has been translated, the following additional information is given: Evan Morgan, *Tao the Great Luminant; essays from Huai-nan tzu* (Shanghai, 1933. Taiwan reprint, 1966), cited as Morgan. Benjamin Wallacker, *The Huai-nan tzu, Book Eleven: Behavior, Culture and the Cosmos* American Oriental Series, vol. 48 (New Haven, 1962), cited as Wallacker. Wm. Theodore De Bary, ed., *Sources of Chinese Tradition* (New York, 1946), cited as *Sources*. In some cases the existing translations have been modified for the sake of conformity.

/1/   This comprehensive outlook is also seen in the commentaries to the *I Ching* 易經 , especially the *Hsi tz'u chuan* 繫辭傳 See Fung Yu-lan, *The Spirit of Chinese Philosophy*, trans. by E. R. Hughes (Boston, 1962), p. 102, where he says, "The *Amplifications* (commentaries) . . . seem to regard the *I* as enabling man to have a positive knowledge; as if, by the method of observing and examining, man is enabled positively to have an all-embracing knowledge of all phenomena."

/2/   I have considered the *HNT* to be a complete text edited by Liu An whose goal apparently was a reconciliation between Taoism and Confuciansim;

thus in many places the text does not appear consistent. See the recent article by Hsü Fu-kuan 許福廣 "Huai nan-tzu yü Liu An te shih-tai" 淮南子於劉安時代 in *Liang-Han ssu-hsiang shih* 兩漢思想史 (Hong Kong, 1975), pp. 73–175, especially the conclusion (pp. 169–70) for a thorough discussion of this position.

/3/ Liu An was a grandson of Emperor Kao-tsu 高祖 (r. 206–194). His kingdom of Huai-nan 淮南 was located in the present day province of Anhui 安徽 . On the scholars congregating at the court of Liu An see Kanaya Osamu 金谷治 , *Shin-kan shisoshi, kenku* 秦漢思想史研究 (A Study of the History of Thought in the Ch'in-Han Period) (Tokyo; Nihon Gakujutsu Shinkokai, 1960), ch. 5. See the review of this work by Ogura Yoshihiko in *Monumenta Serica* 12 (1962): 405–9.

/4/ In 122 B.C., Liu An was indicted on conspiracy charges and was to be executed. However, Liu committed suicide before the sentence was carried out. For his biography see *Shih-chi* 118 and *Han shu* 44. For the theory that Emperor Wu was jealous of Liu An's court, see Ch'ien Mu 錢穆 *Ch'in Han shih* 秦汉史 (Hong Kong, 1957), pp. 70–71. For the life of Liu An and the various political intrigues surrounding him see Benjamin Wallacker, "Liu An, Second King of Huai-nan (180?–122 B.C.)" in *JAOS* 92.

/5/ See the preface of Kao Yu, translated in Wallacker, p. 5.

/6/ See Wallacker, *JAOS* 92: 32, n. 10.

/7/ Liu Hsiang's biography in *Han shu*, ch. 36. I-wen yin-shu kuan reprint of Wang Hsien-ch'ien 王先謙 *Han-shu pu-chu* 漢書補注 36/6b.

/8/ Apparently the two commentaries became mixed when the text was re-edited during the Sung. T'ao Fang-ch'i 陶方琦 (1845–74) in his *Han tzu shih wen ch'ao* 漢孳室文鈔 contained in the *Shao-hsing hsien-cheng i-shu* 紹興先正遺書 1887, established which chapters have commentary by Hsu, and which by Kao. For a textual history of *HNT* see Wu Tsu-yu 吳則虞 , "Huai-nan-tzu shu-lu" 淮南子書錄 in *Wen Shih* 2 (1963): 271–315. For a discussion of the political theories in *HNT* see Hu Shih 胡適 , *Huai-nan wang shu* 淮南王書 (Shanghai: Hsin-yueh shu-chu, 1931). For a review of recent scholarship on *HNT* see the article by Yu Ta-ch'eng 于大成 . "Liu-shih nien-lai chih Huai-nan-tzu hsuen" 六十年來之淮南子學 , in *Liu-shih nien-lai chih kuo-hsueh* 六十年來之國學 , 5 vols. (Taipei, 1972–74), 507–56.

/9/ Translation of D. C. Lao from his *Lao tzu Tao Te Ching* (Baltimore: Penguin Books, 1963), p. 108.

/10/ Donald Munro discusses the distinctions between the Taoist and the Confucian and says one of the attributes of the Taoist is his desire to rid himself of the evaluating mind. Donald Munro, *The Concept of Man in Early China* (Stanford, 1969), p. 11.

/11/   *Chuang-tzu chi-shih* 莊子集釋 , ed. Kuo Ch'ing-fan 郭慶藩 (Taipei, 1967), chapter 2, p. 35; originally published 1894. Translation from Burton Watson, *Chuang Tzu: Basic Writings* (New York, 1964), p. 36.

/12/   *Chuang-tzu chi-shih*, chapter 2, pp. 34–35. Translation from Watson, *Chuang Tzu: Basic Writings*, p. 36.

/13/   *Chuang-tzu chi-shih*, chapter 27, pp. 265–65. Watson, *Chuang Tzu: Basic Writings*, pp. 107–8.

/14/   *Shih t'ung t'ung-shih* 史通通釋 , ed. P'u Ch'i-lung 浦起龍 and originally published c. 1971. I have used the edition published in Hong Kong by T'ai-p'ing shu-chu, 1964, p. 93 of the 2nd section (ch. 36; autobiography).

/15/   See Evan Morgan, "The Operations and Manifestations of the Tao Exemplified in History," in *JNCBRAS* 52 (1921); 1–39, where he numbers each illustration in his translation of chapter 12 of *HNT*.

/16/   Duke Mu of Ch'in (r. 659–21 B.C.), against all advice from his ministers, mobilized his troops and advanced against the state of Cheng 鄭 , intending to attack them by surprise. A merchant of Cheng encountered the Ch'in troops on their way to his state and feigned orders from the Cheng ruler to the effect that he was to present 12 head of cattle to the Ch'in ruler for a feast. This gave Cheng and her allies enough time to outflank Ch'in and defeat her. Thus the *HNT* sums up: "Thus it comes to pass; sometimes in certain things, fidelity becomes a fault and the telling of an untruth a great merit" (*HNHL* 13/15b; Morgan, p. 161). Similar allegorical tales are told of Su Ch'in (d. 312 B.C.) and Kuan Tzu, who became Ch'i's prime minister in 684 B.C.

/17/   *Hsün-tzu chi-chieh* 荀子集解 , ed. Wang Hsien-ch'ien (Shanghai: Commercial Press, Kuo-hsüeh chi-pen ts'ung-shu edition), section 2, p. 7. Translation of Homer Dubs, *The Works of Hsüntzu* (London: Arthur Probsthain, 1928; Taiwan reprint, 1966), p. 74. See also Fung Yu-ian, *A History of Chinese Philosophy* (trans. Derk Bodde), Princeton: Princeton University Press, 1952, vol. I, pp. 282–84, who describes Hsün Tzu's universe as "static."

/18/   See Joseph Needham, et al., *Science and Civilisation in China*, (Cambridge: Cambridge University Press, 1959), 3:642, for a translation and discussion of this section.

/19/   See especially those given in chapters 2 (*HNHL* 2/1a–2a—which is similar to that in chapter 2 of *Chuang-tzu*), 3 (*HNHL* 3/1–3b) and 7 (*HNHL* 7/1a–1b).

/20/   Isaiah Berlin, *The Hedgehog and the Fox; An Essay on Tolstoy's View of History* (New York: Mentor Books, 1957) p. 7.

/21/   *Ibid.*, p. 8. Dante, Plato, Lucretius, Pascal, Hegel, Dostoevsky, Nietzsche, Ibsen, and Proust are, for Berlin, hedgehogs, while Shakespeare, Herodotus, Aristotle, Montaigne, Erasmus, Molière, Goethe, Pushkin, Balzac, and Joyce represent the foxes.

/22/   *Wen-shih t'ung-i* 文史通義 , first published in 1833. I have used the T'ai-p'ing shu-chu edition published in Hong Kong in 1964. The analogy occurs in the section 3 of the chapter "Shu Chiao" 書教 (*The Teaching of History*), p. 13.

/23/   Davis S. Nivison, *The Life and Thought of Chang Hsüeh-ch'eng* (*1738–1801*) (Stanford, 1966), p. 13.

/24/   *Wen-shih t'ung-i*, p. 13.

/25/   Nivison, *The Life and Thought of Cang Hsüeh-ch'eng*, p. 222.

/26/   Cheng Ch'iao appears to have anticipated me here; see the general preface to his *T'ung Chih* 通志 , part of which is translated in *Sources*, pp. 42–44. Cheng says, "Ssu-ma Ch'ien is to Pan Ku as a dragon is to a pig" (p. 444). This, of course, does not represent fox and hedgehog thinking exactly as we have been discussing it here. Cheng calls Pan Ku a pig because he felt that Pan, in not carrying on the tradition of the *Shih-chi*, had done a tremendous disservice to Chinese historiography.

The Five Phases, Magic Squares, and Schematic Cosmography

John S. Major

The eclectic Taoist compendium known as the *Huai-nan-tzu* 淮南子 or the *Huai-nan hung-lieh chieh* 淮南鴻烈解 , written at an academy under the patronage of Liu An, Prince of Huai-nan, and presented to the court of Han Wu-ti not later than 139 B.C., contains a chapter, the fourth, entitled *Ti-hsing hsün* 墬形訓 , or "The Treatise on Topography." As a comprehensive work on topography and the place of the earth and its creatures in the cosmos; as a philosophical document of known date, and as a work that continues the theories of the Tsou Yen 騶衍 school of cosmology in relatively pure form, it is of great importance for the history of the Chinese tradition of cosmology and natural science. This paper will deal with a few of the issues of schematic cosmography raised by a study of "The Treatise on Topography" (hereafter referred to as HNT 4).

The implied cosmogony of the chapter is consistent with the famous cosmogonic passage at the beginning of *Huai-nan tzu*, chapter 1, which describes how the Tao divides into *yin* and *yang* 陰陽 , which further divide into the Five Phases (*wu hsing* 五行). From these the "myriad things" come into being by a process of species differentiation of plants, animals, and minerals from first ancestors with Five-Phase correlations. The chapter also mentions the cosmic triad of heaven, earth, and man, and uses the numerology of three and nine to account for the gestation periods of various creatures.

"The Treatise on Topography" is solidly within the Tsou Yen school of cosmology, wherein all phenomena in the universe (whether "organic" or "inorganic" in Western terms) can be analyzed and understood in terms of *yin-yang*/Five Phase categorical reasoning. Its description of the heavens and the earth follows the *kai-t'ien* 開天 model, which holds that the earth is flat or slightly domed and "square" (i.e., defined by the solstitial and equinoctial points projected onto the celestial equator), and that the heavens are flat or domed, parallel to the earth's plane, and "round" (i.e., defined by the circle of the celestial equator). The treatise further follows Tsou Yen's cosmography in treating the earth as divided into nine great continents. It takes one of these "supercontinents" as the *oikoumene*, divided into nine continents, one of which is China. China itself is further schematically divided into nine provinces.

The cosmography is described in a brief passage in HNT 4 as follows:/1/ "The world has nine continents (*chiu chou* 九州) and eight ultimate end-points (of the eight directions). There are nine provinces (*chiu chou*), nine mountains, and nine passes; nine marshes, eight winds, and six rivers." Detailed analysis of this passage shows that the term *chiu chou* can refer variously to the nine "supercontinents" of Tsou Yen's cosmography, to the nine subdivisions or continents of one of those supercontinents (namely the continent *ch'ih hsien* 赤县 , one subdivision of which, *ch'ih-hsien shen-chou* 赤县神州 , is China itself), and to the nine provinces into which Yü the Great divided China after his conquest of the flood. Thus as a first step in studying the schematic cosmography of the *Huai-nan-tzu* and related texts, we must see how these different divisions fit together.

The usual interpretation of Tsou Yen's "great nine continents" theory has the continent which includes China at the center, with the other eight spread out across the waters in the eight cardinal directions./2/ This produces a tidy arrangement—with China in the southeastern corner of the continent, and Mt. K'un-lun 昆侖 (or 崑崙) standing in the center, in fact in the center of the universe/3/—as follows:

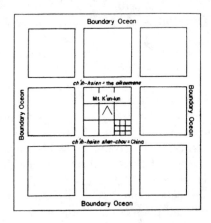

Figure 1. Mt. K'un-lun as the center of the world

The principal difficulty of this conventional mapping is that the *kai-t'ien* cosmology, like virtually any other early cosmology, requires for completeness that the center of the universe be oriented to some astronomical fixed point; and the Chinese knew full well that K'un-lun, as they located it, did not correspond to either of the logical celestial midpoints: the North Pole and the equator. (This in no way denies the identification

of K'un-lun with Mt. Meru, the central pillar, etc.; it merely points out
that the cosmology contains contradictions that cannot be solved logically. As long as the "World Mountain" is given a specific earthly location [except if it were the North Pole] it cannot correspond exactly to the
astronomical reference point; but it can still embody the concept of the
unrotating pivot. In fact there are at least five terrestrial mountains
named K'un-lun by the Chinese, but we are dealing here with a cosmic,
not a physical, mountain.)

What then are the implications for the mapping of the nine continents of taking a celestial reference point for the center of the universe?
Looking first at the polar axis/equatorial plane model of the *kai-t'ien*
cosmology that produces the best astronomical results, we get the following map:

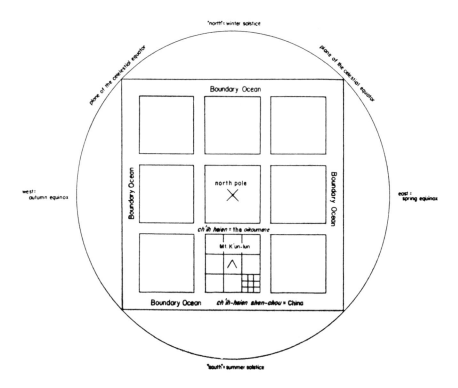

Figure 2. The North Pole as the center of the world

This scheme does not do away with all difficulties, for in order for
China to be south of the North Pole, *ch'ih hsien*—the *oikoumene*—must
occupy one of the lower three positions. Since China is in the southeast
corner of the *oikoumene*, that requirement places China in the far south,

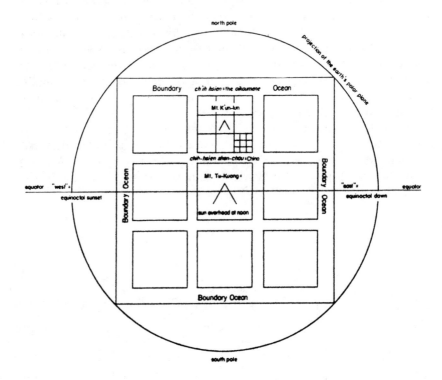

Figure 3. The sun at the equator at noon at the equinoxes as the center
of the world

virtually at the equator—obviously unsatisfactory in terms of terrestrial
geography.

Let us look, then, at what happens when we select the other candi-
date for the celestial reference point: the equator (or, more exactly, the
point, marked by Mt. Tu-kuang 都廣 , directly overhead at the equator occupied by the sun at noon at the equinoxes, by implication taking a
point somewhat to the west of China as approaching 0° longitude.)

This equatorial axis/polar plane model provides a much better
accommodation of terrestrial geography than does figure 2, but at the
cost of putting the earth in the "wrong" celestial plane. For most purposes, as stated earlier, the *kai-t'ien* cosmology takes the plane of the
earth to be a projection into space of the equator, with "corners" at the
equinoxes and solstices; and not as a polar plane, as here.

The conclusion of this small excursion into cosmological mapping
must be that while the several possible midpoints of the universe: the
center of China, Mt. K'un-lun, Mt. Tu-kuang at the equator, and the

north pole, all play the same conceptual role in the cosmology, and all have validity for some purposes, it is impossible to draw a map in such a way that the four coincide. That does not per se render the conception ridiculous; it does mean that we must not approach it with too literal a mind.

Having followed the various types of nine continents to the edges of the earth and beyond, let us return to our starting point and consider the *chiu chou* (nine provinces, this time) of HNT 4, and some of their early analogues.

The earliest system of nine provinces, and the presumed model for all later ones, is of course found in the "Yü kung" 禹貢 chapter of the *Shu ching*. That chapter purports to follow the path of Yü around China, and each province, with its mountains, rivers, soils, tributes, etc., is described in turn. The treatment may be said to be more naturalistic than tabular or schematic. It is interesting to compare this approach with that of three other systems of nine provinces: those of *Chou li*, chapter 33, *Lü-shih ch'un-ch'iu* (*Lscc*), chapter 13, and *Huai-nan-tzu*, chapter 4. If one accepts a Warring States date for the basic content, though not necessarily the specific wording, of the *Chou li*, the four works are successively later in date; reflecting that fact, they tend to become successively schematic and tabular in approach. For example, the *Chou li* account still locates the provinces partly with reference to natural landmarks: *Lscc* locates them with reference to states of the Warring States period; the list in HNT 4 is in contrast completely schematic, locating the provinces simply in terms of the eight cardinal directions plus the center.

The names of the nine provinces in the "Yü kung," *Chou li*, and *Lscc* show some points of difference but are generally compatible with one another. The names of the nine provinces in HNT 4 are markedly different. The reason for this is not known for certain. It has been suggested/4/ that the nine provinces of HNT 4 derive more directly from the Tsou Yen tradition than from the "Yü kung," etc., and that may relate to this considerable difference in nomenclature. In addition, HNT 4 differs from the other three in giving each province an alternate name or "attribute name": "land of agriculture," etc.

The above mentioned points will be clarified by having reference to Table 1, a direct comparison of the nine provinces in these four works.

It will be noted that in no case do the province names in HNT 4 have the same name for the same direction in common with the lists of provinces in the other three works under study. The name Chi 冀 is shared, but in "Yü kung," *Chou li*, and *Lscc* it indicates a northern location, while in HNT 4 it is the name of the center. The western province in HNT 4 is Yen 弇 , similar visually and phonetically to the Yen 兗 east-central province of the other three. The Morohashi *Dai*

TABLE 1: THE NAMES AND LOCATIONS OF THE NINE PROVINCES

Name of Book	Province	Location	Province	Location	Province	Location	Province	Location	Province	Location
SHU CHING III:1 Yü Kung	Chi 冀	[north]	Yen 兗	Between the Chi 河 and the Yellow River [northeast]	Ch'ing 青	Bounded by the sea and Mt. Tai 岱 [east]	Hsü 徐	Bounded by the Sea, Mt. Tai 岱 and the Huai 淮 River [east central]	Yang 揚	Bounded by the sea and the Huai 淮 R. [Southeast]
CHOU LI 33	Chi (8)	The inner part of the Yellow River 河內 [north central]	Yen (5)	East of the Yellow River 河東 [east-central]	Ch'ing (4)	East	P'ing (9) 并	North	Yang (1)	Southeast
LÜ SHIH CH'UN CH'IU 13	Chi (2)	Between the two [branches of the] Yellow River 兩河之內 [north]	Yen (3) [= wei 渦]	Between the Chi 濟 and the Yellow River [east-central]	Ch'ing (4) [= Chi 齊]	East	Hsü (5) [= Lu 魯]	North of the Szu 泗 River 泗上 [east central]	Yang (6) [= Yüeh 越]	Southeast
HUAI NAN TZU 4	Ch'i (7) 沇 [= Ch'êng Tu 成土]	North	Po (8) 薄 [= yin t'u 隱土]	Northeast	Yang (9) 陽 [= Shen Tu 申土]	East	T'ai (6) 台 [= Fei Tu 肥土]	Northwest	Shen (1) 神 [= nung t'u 農土]	Southeast

Province	Location	Province	Location	Province	Location	Province	Location
Ching 荆	Bounded by Mt. Ching 荆 and the South of Mt. Heng 衡 [South]	Yü 豫 (3)	Bounded by Mt. Ching 荆 and the Yellow River [Center]	Liang 梁	Bordered by the South of Mt. Hua 華 and the Black River [Southwest]	Yung 雍 (6)	Between the Black River 黑水 and the Western Yellow River 西河 [West]
Ching " (2)	South	Yü " (7)	South of the Yellow River [Center]	Yu 幽 (7)	Northeast	Yung "	West
Ching " (7) |= Chu 楚	South	Yü " (1) |= Chou 國	Between the Yellow River and the Han 漢 R. [Center]	Yu " (9) |= Yen 燕	North	Yung " (8) |= Ch'in 秦	West
Tz'u 次 (2) |= Wu Tu 天土	South	Chi 冀 (5) |= Chung Tu 中土	Center	Jung 戎 (3) |= Tao Tu 㐱土	Southwest	Yen 弁 (4) |= P'ing Tu 枰土	West

(Note: In Table 1, the numbers in parentheses refer to the original order of enumeration of the nine provinces; the order has been transposed in the table to facilitate comparison of parallel nomenclature. Directions given in square brackets do not appear in the text, but indicate the directional relationship of the given province to others in the same text.)

Kanwa jiten does not indicate any relationship between the two characters, but the similarity is great enough to cause one to suspect an original identity and a later copyist's error in HNT. But again, the direction is different. Yang 陽 , the eastern province in HNT 4, is perhaps related to Yang 楊 , the southeastern province in the other three texts. But the use of Yang 陽 for east, the direction of the sunrise probably is dictated in HNT on cosmological grounds, and so the likelihood of its being related to Yang 楊 , apparently an old place-name in southeastern China, is diminished. Finally, the "attribute name" of Yen, the western province in HNT 4, is P'ing 幷 , identical to the name of the northern province in the *Chou li* list.

It is clear, then, that the province-names in "Yü kung" *Chou li*, and *Lscc* are part of a single tradition, one that is shared hardly at all by HNT 4. The listing of names in HNT 4, furthermore, is done purely by compass points, without reference to landmarks or political divisions, and is done in a strict clockwise serial order: southeast, south, southwest, west, center, northwest, north, northeast, east. Such precision seems to have little to do with the actual, irregular terrestrial world. One begins to ask, what after all are these nine provinces in the *Huai-nan-tzu*? Can they be provinces of China, or is Ku Chieh-kang correct in suggesting that they are the nine subcontinents of Tsou Yen's *ch'ih hsien* continent, with *shen chou*, China, in the southwest? (See the "Textual Note" at the end of this article.) Perhaps these names do represent a survival of the actual nomenclature of the Tsou Yen school; unfortunately, there is no way to tell for sure.

In any case, the nine provinces of HNT 4 certainly bring geography to a level of systematization that divorces it almost completely from the area it is describing, whatever that area is; to draw a "map" of HNT's nine provinces, one does not need a map of any earthly region at all. The "map" is simply the basic grid pattern underlying any division of an area into nine equal parts:

NW T'ai 台	N Ch'i 沘	NE Po 薄
W Yen 弇	C Chi 冀	E Yang 陽
SW Jung 戎	S Tz'u 次	SE Shen 神

Figure 4. The Nine Provinces in *Huai-nan-tzu* 4

This grid pattern, however, is by no means the only purely system-atic geography to be found in the literature with which we have been dealing. The last stage in this brief journey through the mazes of geo-graphical systems in ancient China must be a comparison of those sys-tems and an examination of certain other analogous ones in the literature of the time.

Leaving for the moment our grid of nine squares (although we shall return to it shortly), let us look at another, and quite different, type of systematic geography. In HNT 4, beginning on page 4b, is a long pas-sage that describes the areal extent of the world: "The borders of each of the nine provinces encompass an area of 1000 *li*." The system implied here is some kind of map or nest of concentric layers radiating out from a central territory. Like much of early geographical thought in China, it has its first manifestation in the "Yü kung," and can be seen in several variations in subsequent works.

Needham has written on this subject, "It is usual to hold that the Yü kung implies a naive map of concentric squares. . . . There is nothing in the text, however, to justify [that view]; this was probably just assumed on the basis of the cosmological doctrine of the square earth. The point is more important than it may seem, for if the zones were thought of as concentric circles, this ancient gradient system might have been one of the sources of the East Asian discoidal tradition of 'religious cosmography'. . . . On the other hand concentric squares would fore-shadow a rectangular grid."/5/ This summary is all right as far as it goes, but there is more to say on the subject, and perhaps there are more substantial reasons than an assumption "on the basis of the cosmological doctrine of the square earth" for adhering to the traditional scheme of concentric squares rather than circles.

There are two traditional interpretations of the territories radiating outward from the royal domain, as found in the "Yü kung." One is as shown in figure 5.

This interpretation takes the royal domain (*ti tu* 帝都) as separate from the named radiating squares. In the other interpretation,/6/ the royal domain is synonymous with the *t'ien* 田 , producing a nest of five rather than six squares./7/

A somewhat similar, but much more elaborate, system of concentric squares (ten in all) emerges from the description of successive outlying regions in *Chou li*, chapter 33./8/ while still another variation, this time with seven concentric squares, is found in the "Chou yü" chapters of the *Kuo yü*./9/

The concentric-squares system of the HNT "Treatise on Topogra-phy" is, not untypically, related to but somewhat different from the others discussed above. It is not, strictly speaking, a system of concentric squares, but a system of discrete squares radiating in the eight cardinal

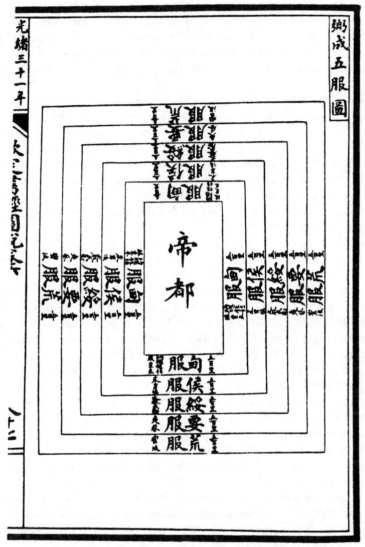

Figure 5. The concentric geography of the "Yü kung"/10/
Needham comments: "Proceeding outwards from the metropolitan
area, we have, in concentric rectangles, (a) the royal domains, (b)
the lands of the tributary feudal princes and lords, (c) the 'zone of
pacification,' i.e., the marshes, where Chinese civilization was in
course of adoption, (d) the zone of allied barbarians, (e) the zone of
cultureless savagery. The systemisation can never have been more
than schematic but Egypt and Rome might have used a similar
image, unconscious of the equally civilised empire at the eastern end
of the world."

directions. This clearly is required by the text, which describes each of the outlying areas "encompassing 1000 *li*"; concentric squares by definition increase in area as they radiate outward, and so will not do here. The map of HNT 4:4b then is as follows.

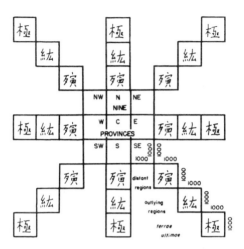

Figure 6. The concentric geography of *Huai-nan-tzu* 4

Despite the numerous points of difference between the map in figure 5 and its variants and that in figure 6, it is clear that they are all of one class. Two further questions may now be raised. First, are the nine-provinces and concentric-squares systematic geographies completely different, and if so, are they incompatible? Second, do these schematic representations of geography relate only to the history of geography and cartography, or are they related to other attempts to apply systems to the cosmos, so that they too take on broader cosmological significance?

In answering the first question, if one were to look only at the "Yü kung," one would probably answer in the affirmative to both parts of the question: the nine provinces and the concentric squares seem not only completely different, but incompatible. That is because, as mentioned earlier, the nine provinces of the "Yü kung" cannot yet be said to be schematic geography. They follow natural boundaries, and the descriptions more-or-less fit the actual geography of China. The concentric squares, on the other hand, already represent a complete systematization of (perhaps) a geographical or ethnographical reality. The two thus fit badly together.

That situation changes completely, however, by the time of the writing of the *Huai-nan-tzu*. The nine provinces and the concentric squares become neither terribly different nor, certainly, incompatible. A 3x3 grid, as in the nine provinces, does after all also define the most basic concentric square figure:

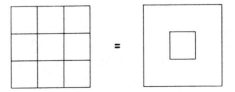

Figure 7. A 3x3 grid defines two concentric squares

In the same way, the nine subcontinents of one of Tsou Yen's continents, if further subdivided into provinces, could with a slight change in point of view be regarded as a simple 9x9 grid of 81 squares, which in turn defines a nest of five concentric squares:

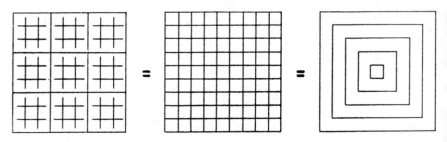

Figure 8. A 9x9 grid defines five concentric squares

If, having gone through this exercise, one looks again at the HNT 4 map in figure 6, it is immediately apparent that it represents a combination of the nine provinces and the concentric squares systems; and further, that it can readily be superimposed on the 9x9 grid of figure 8.

The "distant regions," "outlying regions," and *terrae ultimae* thus begin to take on more meaning. It is possible, of course, that the HNT 4 map in figure 6 has no reference beyond the culture-area of China, that it simply follows the "Yü kung" concentric-squares tradition of giving a schematic map of successively more barbarous regions beyond the area of Chinese rule. But if one were to follow logically the clearly schematic implications of this "geography" that has no specific reference to the known geography of any region, one steps into the realm of pure cosmographical theory of the Tsou Yen school. If one were to take the *chiu chou* of this section to be at the center of one of Tsou Yen's great continents, then the "distant regions" all would lie in the concentric square just beyond their borders: the "outlying regions" would be in the next concentric square, and the *terrae ultimae*, true to the name, would be in the outermost square. Put differently, those distant, outlying, and ultimate regions *define* the nine provinces as being in the central 3x3 grid, and also define the outer three concentric squares and the shape and

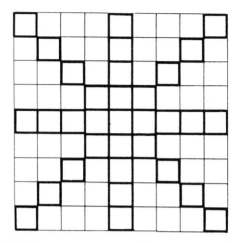

Figure 9. The HNT 4 map superimposed on a 9x9 grid

disposition of the eight outer continents. That solves the question of the odd-looking pattern of squares radiating off in the eight directions: they are all that are needed to define the three concentric squares (whereas if this map were following the "Yü kung" tradition of concentric squares of successively barbarous regions, the central provinces would simply be surrounded by continuous rings 2,000 *li* wide, rather than by areas defined as eight "distant regions," etc., each 1,000 *li* square). Solved too is the problem of the meaning of *chi* 冀 : here it certainly does not have the meaning of "the outermost extension of the four directions"—i.e., the equinoctial and solstitial points on the celestial equator, as it does in the passage quoted at the beginning of this paper, but rather is the outer rim of one of the great continents, viewed from the center.

The following conclusions can be stated at this point: First, the schematic map of HNT 4:4b is compatible in general tone, and does not conflict in details, with the earlier schematic and realistic geographical sections of the chapter. Second, the map is derived from and is a synthesis of both the nine provinces and concentric-squares geographical traditions, and represents a significant theoretical advance over earlier forms of those systems in depicting a comprehensive view of the configuration of the terrestrial surface. Third, the map and the system it represents are completely in harmony with what is known of Tsou Yen's cosmography, and therefore probably is an authentic teaching of the Tsou Yen school, preserved in technical and schematic language that has hitherto prevented it from being fully understood.

Still unanswered is the second question asked above: are the systems of the nine provinces and the concentric squares related to other systematic representations in ancient Chinese literature, so that they have significance beyond geography and cartography? The answer to this question must be an emphatic "yes."

Literature of the Warring States period refers often to two writings or diagrams of a magical and symbolical nature, the *Lo shu* 姜書 (Lo Writing) and the *Ho t'u* 河圖 (River Chart) (see figures 10 & 14). For example, the *Great Treatise*, "Ta-chuan" 大傳 , of the *I Ching* says, "The Yellow River brought forth a map and the Lo River brought forth a writing; the holy men took these as models."/11/ The Lo Writing was supposed to have appeared on the back of a sacred tortoise (as a 3x3 grid!) and has from the first been associated with numerology, and particularly with that of the "Hung fan" 洪範 chapter of the *Shu ching*. The Lo Writing (at least in the form of the diagram associated with Chu Hsi, and accepted as definitive since his time; see figure 10) is the magic square of three, a Chinese mathematical invention./12/ (The earliest indisputable magic square known anywhere is contained in the description of the *ming-t'ang* 明堂 "cosmological hall," in the chapter of that name [chapter 69] in *Ta-tai li-chi*; that configuration is derived from the Lo Writing.) The River Chart was supposed to have emerged from the Yellow River and is also numerological in character, associated not with nine but with ten. Rickett has shown/13/ that it is related to calendar charts of the type best known from the "Yüeh ling" 月令 (*Li chi* chapter 6, derived from portions of *Lscc* chapters 1–12), but also in *Kuan-tzu* III, 8,/14/ HNT 5, and even, apparently, in the text of the Ch'u Silk Manuscript./15/ Moreover, both the Lo Writing and the River Chart appear to be related to the systematic geographies of the nine provinces and the concentric squares.

That the Lo Writing forms a 3x3 grid immediately suggests the nine provinces, but does not prove the point; such a grid after all is a common enough geometrical figure. Knowledge that the Lo Writing is a magic square seems, in fact, at first glance to lead in the opposite direction. It will be remembered that the nine provinces in HNT 4 were listed in the following order:

NW	N	NE
6	7	8
W	C	E
4	5	9
SW	S	SE
3	2	1

which looks nothing like the arrangement of the numbers in the magic square of three as they appear in the Lo Writing:

4	9	2
3	5	7
8	1	6

If, using the standard numerical correlates of the Five Phases (from the "Hung fan" chapter of the *Shu ching*: 1,6 = water, 2,7 = fire, 3,8 = wood, 4,9 = metal, 5,10 = earth), one replaces the numbers in the magic square with the Phases:

M (4)	M (9)	F (2)
Wo. (3)	E (5)	F (7)
Wo. (8)	Wa. (1)	Wa. (6)

the situation hardly seems improved; the directional correlates, with the exception of earth at the center, are all wrong. A second look, however, shows that we are not so far astray. It will be remembered that the mutual production cycle of the Phases (earth produces metal, which melts to produce water, which nourishes to produce wood, etc.) can be used to account for cyclical phenomena such as the seasons, and can be graphed as shown in figure 11.

A slight alteration of this graph can be made to include the "latent" or "undeveloped" forms of the Phases as shown in figure 12./16/ Looking

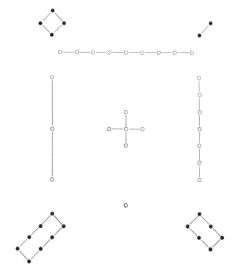

Figure 10. Chu Hsi's Lo Writing

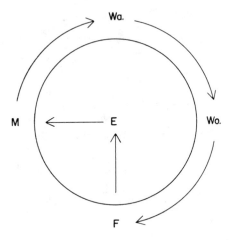

Figure 11. The cycle of the mutual production order of the Five Phases

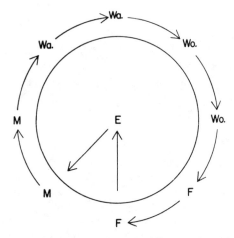

Figure 12. The cycle of the mutual production order of the Five Phases,
           expanded

again at the Phase-correlations of the magic square, it will be noticed
that the Phases, though not correlated to the directions, do form an
order: the mutual overcoming order, i.e., the converse of the mutual
production order: fire overcomes metal, which overcomes wood, etc. The
mutual overcoming order is a cyclical one that can be graphed in exactly
the same form as the mutual production order in figure 12. It thus
becomes obvious that the number-correlates of the Five Phases, arranged
in a magic square of three, and the Nine (the basic Five plus the
coming-into-being forms in the four corners) Phases in the mutual

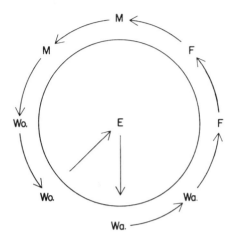

Figure 13.The cycle of the mutual overcoming order of the Five
        Phases, expanded

overcoming order when depicted in a 3x3 grid, are identical and freely
interchangeable:

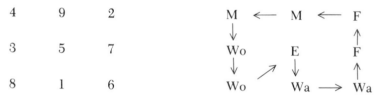

The magic square of three was discovered in China during the Warring States period, if not earlier; the Five Phases correlates of the numbers
of the magic square of three form the mutual overcoming order of the
Phases. That is unlikely to be a fortuitous coincidence; more probably it is
an example of the high degree of integration characteristic of ancient
Chinese cosmology. In Tsou Yen and the members of his school we are not
dealing with some quaint Oriental gentlemen who had rather odd notions
of how the universe was put together; rather they were serious thinkers
committed to an integrated, organic cosmology, who would have sought
confirmation for their views in any and all correspondences of cosmography, numerology, and natural phenomena./17/ Thus we may reasonably conclude that the figure of the 3x3 grid was widely used in the
schematic cosmography of the Tsou Yen school, that both the Lo Writing
and the schematic 3x3 map of the nine provinces are manifestations of the
same type of reasoning, that the Five Phase correlates of those figures were
likely to have been known and regarded as significant, and that all of these

things taken together provide evidence for a truly remarkable degree of integration in the cosmological speculations of ancient China.

What, then, of the River Chart, the other famous cosmological diagram of the Chou period? As with the Lo Writing, the earliest pictorial representation that we have of it is that of Chu Hsi, but such a late date is not too surprising. Cammann points out/18/ that the Lo Writing magic square was a carefully-guarded esoteric secret even in T'ang times, and the same would be true of a schematic drawing of the River Chart. Rickett gives evidence that Chu Hsi's diagram may far predate the Sung, and may even be traced as far back as the Han./19/

Chu Hsi's River Chart is as follows:

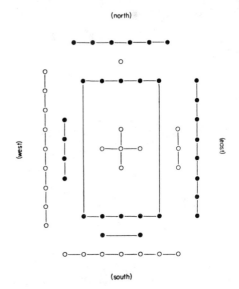

Figure 14. Chu Hsi's River Chart

Rickett points out the similarity between this graph and the calendar charts of the "Yu-kuan" chapter of the *Kuan-tzu*, the "Yüeh ling" chapter of the *Li chi*, etc./20/ Kuo Mo-jo has reconstructed the Yu-kuan chart in a way that shows this similarity very strikingly.

It should be noted in passing that these calendar charts are not simply calendars, but are full of prognostications, lucky and unlucky days, and Five Phase correlations, so that their association with the *yin-yang*/ Five Phase school of cosmology is beyond question.

The similarity of the calendar charts to the systems of concentric-square geography in turn is striking. If one redraws the Yu-kuan chart, in a way that does not alter the data but does change its visual impression somewhat, the resemblance to a system of concentric squares is remarkable.

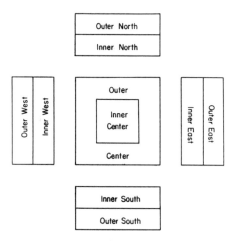

Figure 15. Kuo Mo-jo's reconstruction of the "Yu-kuan" calendar chart/21/

The final stage of analysis of the River Chart will illustrate the extent to which this ancient symbolic figure, like the Lo Writing, integrates numerology, the Five Phases, and the schematic organization of space.

The numbers of the River Chart, as seen in Chu Hsi's diagram (fig. 14) are as follows:

$$6$$
$$1$$
$$9 \qquad 4 \qquad 10/5 \qquad 3 \qquad 8$$
$$2$$
$$7$$

In other words, each of the five directions contains both the *yin* (even) and *yang* (odd) number-correlates for that direction. The River Chart can also be arrayed in a 3x3 grid, with the *yin* numbers at the corners (as in the case of the Lo Writing), the odd numbers in the four cardinal directions, and with 5 and 10 (*yang* and *yin* earth) in the center:

6	1	8
9	10/5	3
4	7	2

The Five Phase directional correlates of these numbers, by definition, form a 3x3 grid of the eight cardinal directions plus the center; the Five Phase correlates of the numbers form the mutual production order of the Phases! (See figs. 11 and 12.)

NW	N	NE
W	C	E
SW	S	SE

```
Wa ——> Wa ——> Wo
 ↑              ↓
 M      E       Wo
 ↑    ↙         ↓
 M      F  <——  F
```

Thus it is apparent that, while both the Lo Writing and the River Chart were used for various purposes and at different levels of sophistication throughout the history of Chinese philosophy and religion,/22/ on perhaps their deepest level of meaning the two figures embody the two prime enumeration orders of the Five Phases arrayed in a schematic organization of space as a 3x3 grid in such a way as to make possible a wide range of numerological and cosmographical manipulations.

This solution to the problem of the Lo Writing and the River Chart seems to show beyond doubt that the grid figure was integral to the developed phase of Chinese cosmology in ways that go far beyond schematic cosmography itself; it provides a key to the understanding of the entire system. That being so, it is worthwhile to take a fresh look at a few other grid figures in early Chinese thought.

Mencius's well-field (*ching-t'ien* 井田) idealization of land tenure and land taxation in the Golden Age is, of course, a 3x3 grid./23/ Having the lord's field, worked in common by the eight families of the outer squares, in the center of the figure, it is a low-level example of the integration of the concentric-squares/grid transformation of the schematic organization of space discussed earlier (see figs. 7 and 8). To be sure, this use of the grid figure is relatively unsophisticated, but it does show the extension of the sort of reasoning discussed here beyond the School of Naturalists into the mainstream of Confucianism itself.

The pervasive influence of the grid figure is also seen in the standard

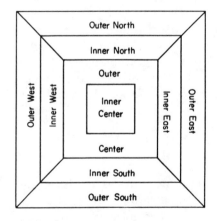

Figure 16. Reconstruction of Yu-kuan calendar chart, redrawn

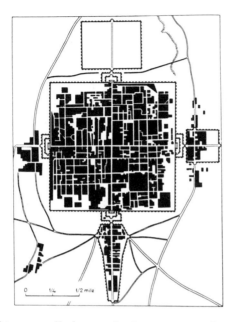

Figure 17. A Chinese walled town built on a typical 2x2 grid plan: Ta-
t'ung-fu, Shansi/24/

Chinese proof of the Pythagorean theorem, as found in the *Chou-pei
suan-ching* 周髀算經 ./25/ The proof is not algebraic, but rather a gen-
eralisation from the case $3^2 + 4^2 = 5^2$: 3x3, 4x4, and 5x5 grids are
drawn adjacent to the sides of a right triangle formed by diagonally
bisecting a rectangle of 3x5 dimension; by counting the resulting squares
of the grids and showing that $9 + 16 = 25$, the theorem is demonstrated.
    Town planning in ancient China also showed a firm commitment to
the grid form. The most simple and most typical form of walled town
had a roughly square or rectangular wall, within which the town was
divided by main streets running east-west and north-south, making a 2x2
grid; the four wards thus formed were further divided by grids of streets
and alleys./26/ At the intersection of the two main streets was a square,
so that the town also showed, in rudimentary form, a merging of the
concentric squares and grid forms of spatial organization (see fig. 17). In
the case of a royal city, the king's palace was not located, as one might
expect, in the central square, but rather in a separate walled square
against the central portion of the northern wall. This was because the
Chinese ruler ritually "faced south"; that is, he symbolically occupied the
position of the Pole Star at the unrotating pivot of the universe./27/ The
king's position thus is analogous to that of the north pole in the cosmic
map depicted in fig. 3. As in that case, the North Pole (the royal

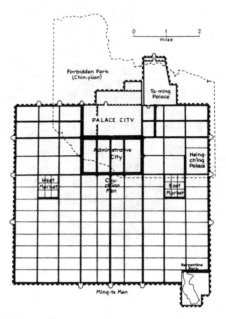

Figure 18. A Chinese Capital City: Ch'ang-an during the T'ang/28/

enclosure) surveys the rest of the world beneath it, while in the city the central crossroads functions as an analog of Mt. Tu-kuang as a ritual "center" for other purposes.

A further hint that Chinese thinkers were aware of the city plan as a grid with significant Five Phase correlations is seen in the Mohist organization of defensive warfare. The defending forces were to include shamans (*wu* 巫) and ether-watchers (*wang-ch'i-che* 望氣者); the ether-watchers were deployed according to a scheme derived from Five Phase numerology—eight eighty-year-old ether-watchers on the eastern wall of the city, etc./29/

Paul Wheatley has shown that the Chinese grid-plan city was, in essence, a type of mandala./30/ A mandala is a symbolic depiction of space in the *Urzeit* of cosmogony; by contemplating, or locating one's self within, the mandala one is able to achieve a correct ritual identification with, or orientation towards, the *axis mundi* itself. The *ming-t'ang*, described in *Ta-Tai li-chi* and constructed by Han Wu-ti, was a mandala based on the magic square of three. A type of tomb found widely throughout China, Korea, and Japan has four walls decorated with paintings of the mythical animals (*ssu ling* 四靈) that symbolize the four Celestial Palaces of the noncircumpolar stars (Caerulean Dragon, east; Vermilion Bird, south; White Tiger, west; Dark Warrior [=Tortoise and serpent], north)./31/ The central space of the tomb then becomes the

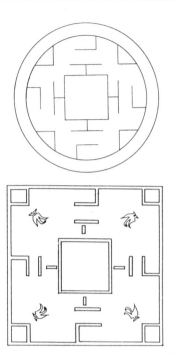

Figure 19. (above) Typical design of Han TLV mirrors
        (below)*Liu-po* game board from Ma-wang-tui Tomb 3/32/

Purple Palace of the circumpolar stars; and the location óf the coffin in
the tomb's center, on the site of the Pole Star itself, transforms the tomb
into a mandala. Our final example of schematic cosmography in this
paper will concern another type of Chinese mandala.

The so-called TLV bronze mirrors of Han date have been studied
intensively, and much is now known about their symbolism./33/ The
TLV motif is found on certain Han sundials, and its significance in that
context has been explored by Needham;/34/ the mirrors are also closely
linked with the TLV playing-board of the game *liu-po* 六博 ./35/ A
pattern similar but not identical to the TLV design is found on the
square "earth plate" of the diviner's board (*shih* 栻 ; the earth plate was
surmounted by a round, movable "heaven dial" marked with the seven
stars of the Dipper; the assembly was used for astronomical direction-
finding by the pointer-stars of the Dipper during daylight hours)./36/

TLV mirrors were produced in abundance in the Han, and frequently bear inscriptions that mark them as auspicious objects; they are
associated particularly with the period of the Hsin dynasty (A.D. 9–23) of
Wang Mang, whose interest in and patronage of *yin-yang*/Five Phase

Figure 20. A bronze TLV mirror from the Hsin Dynasty (The Cull
Collection,

cosmology is well known. While it will not be possible here to review the
manifold discoveries about the cosmic symbolism of the TLV motif
related in the above-mentioned studies, the links between the TLV mirror, the sundial, the diviner's board, and the *liu-po* board (see fig. 19)
require that some attention be paid to possible connections between the
TLV motif and schematic cosmography.

A particularly fine TLV mirror from the Hsin dynasty/37/ provides
a good focus for our study (fig. 20). It has a central boss, an inner square
set with nine smaller bosses, a typical TLV pattern against a background
of scrolls and fabulous birds and animals with eight smaller bosses
distributed between the T's, and four conentric outer rings containing an
inscription, a line-pattern, a zig-zag pattern, and a scroll pattern that

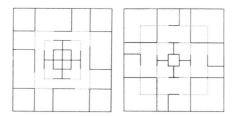

Figure 21.TLV markings on *liu-po* boards define three concentric squares beyond the central square of the *oikoumene*

may represent clouds or waves. The inscription reads:

> The Hsin (Dynasty) has excellent copper, it comes from Tan-yang; Refined and worked with silver and tin, it is clear and bright. The Shang-fang [state workshops] have made (this) mirror, (which) is completely without flaw; To the left the Dragon and to the right the Tiger avert misfortune; the Vermilion Bird and the Dark Warrior [= the tortoise] are in accord with yin and yang. May your sons and grandsons be complete in number and dwell in the center; May you long preserve your two parents in happiness and good fortune. May your longevity be like that of metal and stone; May your lot be that of a prince./38/

Several mirrors closely similar but not completely identical to this one are known. The inscription on one/39/ provides a key to the line "may your sons and grandsons be complete in number," and to the eight smaller bosses (nine including the central knob) on the mirror; the inscription on the similar mirror reads, "may your eight sons and your nine grandsons govern."/40/

One must ask, however, whether the bosses do not have a significance beyond indicating a "complete number" of sons and grandsons. The wish that they "be in the center" provides a clue for the direction our inquiry should take. "To be in the center" may be a quite secular wish for success at the capital, but on the context of an object of known cosmic symbolism, it should mean something more. Getting to the center is the function of a mandala, and that meaning must be implied here as well./41/ If the TLV mirror is a mandala, we must examine it for a symbolic organization of space; once sought, it is easily found.

Recall that in fig. 1 we posited that Mt. K'un-lun could be taken as the center of the universe. If we take the central boss of the mirror to represent K'un-lun, then the central square becomes the central one of the nine continents of Tsou Yen's nine continents theory; the smaller bosses, correlated with the twelve Earthly Branches (chih 支) represent the twelve directions pointed out in succession by the "handle" of The Dipper through the twelve months of the year, thus orienting the central continent

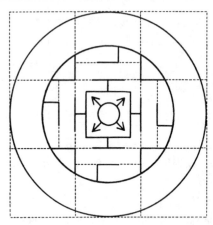

Figure 22. The pattern of a TLV mirror is formed when a *pi* disc, representing heaven, is placed over a *liu-po* board, representing earth. /42/

on both time and space. The eight small bosses interspersed among the T figures indicate that eight other continents lie outside the central one. The central continent is the known world, the *oikoumene*, with the world-pillar in its center; the outer continents then are not on the earth's surface, but in the cosmic space, the realm of fabulous creatures riding amidst the clouds. What then of the TLV markings themselves—do they serve a cosmographic function? If one looks again at the schematic concentric map derived from HNT 4 (fig. 6), it will be clear that they do.

   Liu-po boards are found in two basic types, in which the stems of the L figures are either the same length as or shorter than the legs of the squares in the four corners of the board. Figure 23 shows that in both cases, however, the TLV markings serve to define three concentric squares beyond the central square of the *oikoumene*, just as the radiating squares of the HNT 4 grid-map in figure 6 define the distant regions, outlying regions, and *terrae ultimae* of Tsou Yen's universe. Figures 8 and 9 showed in turn that the concentric-squares figures of the HNT 4 map could be superimposed on a 9x9 grid, the schematic representation of Tsou Yen's nine continents, to define the boundaries and dimensions of the continents. *Liu-po* boards thus are schematic maps of the universe in the "square" plane of the earth, i.e., the plane of the clelestial equator with "corners" at the equinoctial and solstitial nodes.

   Cammann has shown that if a jade disc of the type known as *pi* 碧 , symbol of the "round" heavens, is placed over a *liu-po* board, symbol of the "square" earth, of the type whereon the stems of the L figures are shorter than the legs of the corner squares, the emergent figure is precisely that of the TLV mirror (see figure 22). The TLV mirror thus is a comprehensive diagram of the cosmos, a mandala; it is round, conforming to the shape of heaven, but the pattern preserves the essential map of the square earth below heaven's dome.

Thus the passage from HNT 4:4b cited above is in effect a description of the cosmic map on the back of a TLV mirror, and the TLV mirror is a description of the cosmos of Tsou Yen's nine continents, with the boundary ocean surrounding the continents in the outermost circle of the mirror, and the mirror's edge representing the edge of the dome of heaven.

In conclusion, one may say that schematic cosmography, embodied in the 3x3 grid and certain closely related figures, seems to have permeated the thinking of the Chinese of the Warring States and Han periods. Philosophers of Tsou Yen's naturalist school were able to use that cosmography to manipulate *yin* and *yang* and the Five Phases to achieve an integrated, intellectually satisfying, and in a special sense even scientific, view of the cosmos.

Textual Note

At the beginning of this article I took a brief passage from HNT 4 as a point of departure to introduce an analysis of the schematic cosmography of Tsou Yen and his school. The text, here presented in a slightly different form, is as follows:/43/ "The world has nine continents and eight ultimate endpoints (of the eight directions). The dry land has nine mountains, the mountains have nine passes. There are nine marshes, eight winds, and six rivers."

The passage closely follows the text of *Lü shih ch'un-ch'iu* 13:1b./44/ Reading the two passages side by side, we find:

Lscc: *T'ien*   *yu*   *chiu*   *yeh,*   *ti*   *yu*   *chiu*   *chou.*   Shang/45/   *yu*
天　有　九　野　地　有　九　州　上　有

chiu   *shan*   *shan*   *yu*   *chiu*   *sai.*   *Tse*   *yu*
九　山　山　有　九　塞　澤　有

HNT: *T'ien*   *ti*   *chih*   *chien,*   *chiu*   *chou*   *pa*   *chi.*   *T'u*   *yu*   *chiu*
天　地　之　間　九　州　八　極　土　有　九

shan   *shan*   *yu*   *chiu*   *sai.*   *Tse*   *yu*
山　山　有　九　塞　澤　有

Lscc: *chiu*   *shu,*   *feng*   *yu*   *pa*   *teng,*   *shui*   *yu*   *liu*   *ch'üan.*
九　數　風　有　八　等　水　有　六　泉

HNT: *chiu*   *shu,*   *feng*   *yu*   *pa*   *teng,*   · *shui*   *yu*   *liu*   *p'in.*
九　數　風　有　八　等　水　有　六　品

The passage presents several difficulties not noted in any commentary that I have seen, nor mentioned by Erkes and R. Wilhelm, respectively, in their studies of HNT 4 and of *Lscc*./46/ The most important of these difficulties is that while the several phrases that make up the passage appear to be exactly parallel, they are not. In the last three phrases, *shu, teng,* and *ch'üan* (or *p'in*) are used as "measures" (numerary adjuncts) for *tse, feng,* and *shui* respectively. (Erkes' attempt at a literal translation, *loc. cit.*, is certainly wrong in this respect:

"unter den Marschen gibt es die neun Sümpfe, unter den Winden gibt es die acht Klassen, und von Gewässern die sechs Rangstufen." So also Wilhelm, *loc. cit.*: "in den Seen gibt es neun Inseln. Vom Wind gibt es acht Arten. Vom Wasser gibt es sechs Ströme.") Thus for parallel construction one would expect *shan* to be a measure for *t'u*, and *sai* a measure for *shan*. But the subsequent detailed listing names of geographical features gives the following categories: *chou, shan, sai, shu, feng,* and *shui.* So the phrases are not parallel in meaning: *shan* and *sai* have the force of "full words" as names of categories; *shu* is first used as a measure for *tse* but then used in place of *tse* in the subsequent enumeration; *teng* and *ch'üan* (*p'in*) are used only as measures.

There are two possible approaches to translating this line. The first, which I have followed above, adheres to the wording as we have it but ignores the seeming parallelism. The second approach assumes that the parallelism should hold throughout, making a literal translation of the text as we now have it unsatisfactory, and therefore that the passage must have been corrupted by either incorrect or dropped characters, or both. (This requires a further assumption that the error was a very early one in the text of *Lscc* 13 before it was copied into HNT 4—a possible victim of Ch'in Shih-Huang-Ti's destruction of books?)

Supposing for the moment that the present text is defective, a possible reconstruction would be the following: "*t'u yu chiu shan* (read *chou*?), *shan yu chiu* [X], *sai* [*yu chiu* Y], *tse yu chiu shu,*" etc. It is not too far-fetched to suppose that the first *shan* is a copyist's error for *chou*. This emendation gains plausibility not only from a slight similarity in form of the two characters, but also from the linking together of *chou* and *t'u*.

This emended reading would yield (beginning with the preceding line): "The world has nine continents and eight ultimate endpoints (of the eight directions). There are nine provinces, nine mountains, and nine passes; nine marshes, eight winds, and six rivers." Thus emended, the passage is a perfect introduction to the lists of provinces, mountains, passes, etc., which follow.

On the other hand, I feel that one normally should be biased against tampering with a received text, although reconstruction of clearly erroneous passages should be attempted when necessary. For that reason I regard my suggested reading as tentative. The degree of supposed error, and hence of necessary reconstruction, is rather large: one must assume first that a flawed passage was slavishly copied from *Lscc* into HNT, and further that this short passage contains one mistaken character (the first *shan* as a mistake for *chou*, as suggested above) and that an additional four characters have dropped out.

This passage contains conceptual problems in addition to the purely textual difficulties just discussed; an enquiry into those conceptual problems can shed additional light on the textual ones as well. One will note that in the emendation of this passage that I have proposed, the word *chou* is used twice in quick succession (*T'ien ti chih chien, chiu chou pa chi. T'u yu chiu chou*), but the meaning in each case is not the same. The word in the first instance must refer to the nine great continents into which "all-under-heaven" (*t'ien ti chih chien = t'ien hsia* 天下) is divided (or perhaps, as we shall see in a moment, the nine subcontinents into which, according to Tsou Yen, each of the nine continents of the world is divided). The word in the second instance refers to the nine provinces into which China itself is divided./47/

At this point, then, we are led into an examination of the whole question of the meaning of *chiu chou*. Ku Chieh-kang/48/ distinguishes three "nine continents" or "nine provinces" theories. The first is the "small nine provinces" theory (*hsiao chiu chou* 小). This derives from the "Yü kung" chapter of the *Shu ching*, and describes, with reasonable plausibility, nine subdivisions of the territory that formed the heartland of China in the Chou period. With reference to this theory, it is clear that the term *chou* is properly translated as "province."

The second theory is the "large nine continents" theory (*ta chiu chou* 大), propounded by Tsou Yen in the third century B.C. This theory held that the entire world is divided into nine great continents, separated from each other by oceans; each of these great continents is further divided into nine subcontinents. China, to which Tsou Yen gave the name "Spiritual Subcontinent of the Vermilion Continent" (*ch'ih-hsien shen-chou* 赤县神州), represented one-ninth of one of the great world-continents, i.e., one of eighty-one such subcontinents. Each of the nine provinces laid out by Yü the Great would then comprise 1/729 of the total area of the world.

The third theory, the "middle nine continents" theory (*chung chiu chou* 中), is also based on Tsou Yen's doctrines. It accepts the division of the world into nine large continents, but focuses its attention on the one large continent that contains China. The other eight apparently are dismissed as unknowable or of scant interest. Ku Chieh-kang contends/49/ that the *chiu chou* of HNT 4:1ab are the nine subcontinents into which this one large continent is divided. The list of *chou* in HNT 4 includes the name of *shen chou* 神州 , which might be equated with, and almost certainly is related to, the *ch'ih-hsien shen chou* of Tsou Yen. The author of HNT 4 places *shen chou* in the southeast, Ku speculates, because it was well known that China is bordered on the south and east by the sea; thus it would be easy to assume that China—*shen chou*—represented the southeast division of a group of nine continents.

Ku's argument that the nine continents here represent the nine subcontinents of Tsou Yen's *ch'ih hsien* continent is reasonable; but it contains an important difficulty. It is that immediately after these subcontinents are given, comes a list of mountains, passes, marshes, etc., that clearly are located in the *provinces* of China itself. A possible explanation for such an abrupt transition is that the identification of China as *shen chou* was so well known from Tsou Yen's theory that our author, writing less than two centuries after Tsou Yen, felt able to shift his frame of reference and abruptly narrow his focus from the subcontinent to the nine provinces, with no loss of clarity for his contemporary readers. But such an explanation is not entirely convincing.

On the other hand, if the emended HNT text is used, we have a reading "There are nine [sub]continents and eight ultimate endpoints. There are nine provinces, nine mountains," etc., which accords perfectly with Ku's "middle *chiu chou* theory"—one can suppose that Tsou Yen's other eight continents are out at the "ultimate endpoints"; that *ch'ih hsien* is divided into nine subcontinents, and that one of these, the "dry land" of China, is further divided into nine provinces. This works beautifully except that by taking the nine provinces given in HNT 4:1b as provinces of China, Ku's argument that *shen chou is* China is destroyed. Thus Ku's theory is upheld by logic but the evidence he adduced is destroyed in the process.

In the end, the case for emending the text in this instance cannot be either accepted or rejected on the basis of Ku Chieh-kang's theory. A reasonable translation emerges from the text as is, which argues for leaving it alone. The emendation that I have discussed has no weight of tradition behind it; nevertheless it produces a somewhat more tidy translation, accords well with Ku Chieh-kang's "middle *chiu chou* theory," and is consistent with the general cosmological framework of the *Huai-nan-tzu* itself./50/ For the time being, the argument must rest at that point.

NOTES

/1/ HNT 4:lab, in the edition of Liu Shu-ya (Liu Wen-tien), *Huai-nan hung-lieh chi-chieh* 淮南鴻烈集解 (Shanghai, Commercial Press, 1926); reprinted, Taipei, 1962. All citations of HNT are from this edition. The passage is here cited and translated in emended form; the case for the proposed emendation is made in detail in the "Textual Note" that concludes this article. There the reader will also find a detailed explanation of how and why the term *chiu chou* can refer to three quite different scales of geographical division in Tsou Yen's schematic cosmography.

/2/ This interpretation is not strictly speaking required by Tsou Yen's nine-provinces theory as it is usually stated, but it is implied by one interpretation of the phrase *chiu-chou pa-chi* 九州八極 in HNT 4:la.

/3/ For a more detailed discussion of the issues raised in this paragraph, see John S. Major, "Topography and Cosmology in Early Han Thought: Chapter Four of the *Huai-nan-tzu*" (Ph.D. diss., Harvard University, 1973), 45–49.

/4/ By Ma Pei-t'ang in "Huai nan chiu-chou ch'ien-shen hou-ying" 淮南九洲前生后影 (The origin and later influence of *Huai-nan-tzu*'s nine provinces), *Yü-kung pan-yüeh k'an* 愚公半月刊 3:1–6 (1934).

/5/ Joseph Needham, *Science and Civilisation in China*, 7 vols. projected (Cambridge, England: At the University Press, 1954–), 3:501–2. Hereafter Needham, *SCC*.

/6/ See the figure in Legge, *Chinese Classics III, Shoo King*, "The Tribute of Yü," p. 147.

/7/ For an analysis of the two interpretations, see Bernard Karlgren, "Glosses on the Book of Documents," *BMFEA* 20:34–315 (1948), pp. 159–60.

/8/ See the figure in Legge, *Shoo King*. "The Tribute of Yü," p. 149.

/9/ See the figure in Karlgren, "Glosses on the Book of Documents," p. 160.

/10/ *Shu-ching t'u-shuo* 書經楚説 chap. 6; reproduced in Needham, *SCC* 3:501; caption from *loc. cit.*

/11/ *I-ching* 7:10a (*SPTK* ed.); trans. R. Wilhelm, *The I Ching or Book of Changes* (trans. from German by C. Baynes; Princeton: Princeton University

Press, 1950), p. 344. For a brief history of the *Lo shu* and the *Ho t'u*, see Needham, *SCC* 3:56–59.

/12/ See S. Cammann, "The Magic Square of Three in Old Chinese Philosophy and Religion," *History of Religions* 1:37–80 (1961), for an account of the early history of the Lo Writing and the magic square.

/13/ W. Allyn Rickett, "An Early Chinese Calendar Chart," *T'oung Pao* 68:195–251 (1960).

/14/ Rickett's article cited in note 13 is a study of the "Yu-kuan" chapter.

/15/ A. F. P. Hulsewé, "Texts in Tombs," *Asiatische Studien/Études Asiatiques* 18/19:78–89 (1965), pp. 83–84.

/16/ The names of the eight *terrae ultimae* emphasize the potentiality of the elements in the northeast, southeast, southwest, and northwest.

/17/ In an earlier version of this paper, I wrote, "In fact, in the present instance it would be tempting to speculate that the magic square of three itself was discovered, not mathematically, but by a process of laying out the mutual overcoming order of Phases in a 3x3 grid pattern and then noticing that their correlate numbers formed a numerologically interesting figure. Such speculation is bolstered by a further numerologically significant correlation: in *yin-yang* theory, even numbers are *yin*, odd numbers are *yang*. In the eight outer squares of the magic square of three, all odd/*yang* numbers are in the 'primary' positions of the four cardinal points, all even/*yin* numbers are in 'secondary' or 'latent' positions at the corners. This does not amount to proof that the magic square was discovered through Five Phase numerological manipulation rather than mathematically, but it is strongly suggestive." Professor Cammann (private correspondence) has convinced me that this speculation is misplaced; it appears certain that the magic square of three was developed in China in very ancient times, and long preceded the correlation of intergers with the Five Phases. Thus the reverse of my speculation is likely to be true—the Five Phases probably took their numerical correlates from the magic square of three. This being so, however, the integration of *yin-yang* and Five Phase numerology surely must have been enhanced by the fortuitous circumstance that the corners of the magic square are all occupied by even/*yin* numbers. It is also significant that in the magic square the numeral nine occupies the square corresponding to due north. The guardian animal of the north (one of the *ssu ling* 四 靈) is the tortoise (encircled by or associated with a snake, together known as the Dark Warrior, *hsüan wu* 玄 武); it was noted above that the Lo Writing, a 3x3 grid, is said to have appeared on the back of a tortoise.

/18/ S. Cammann, "Old Chinese Magic Squares," *Sinologica* 7:14–53 (1963), p. 15; see also S. Cammann, "The Magic Square of Three in Old Chinese Philosophy and Religion," *History of Religions* 1:37–80 (1961).

/19/ Rickett, "An Early Chinese Calendar Chart," p. 205.

/20/ *Ibid.*, pp. 205–7.

/21/   *Ibid.*, 207.

/22/   Michael Saso, "What is the Ho-t'u?," *History of Religions* 17.3 + 4:399–4126 (special issue on Chinese religions, February-May 1978). For other magic-diagram numerological possibilities of the Lo-shu, and especially of the Ho-t'u, see Needham, *SCC* 3:55–59.

/23/   Mencius III.I.3:13–20; James Legge, *The Chinese Classics*, vol. 2, *The Book of Mencius* (Oxford, 1895) 243–45.

/24/   From Sen-dou Chang in Skinner, map 1.

/25/   Needham, *SCC* 3:22–23; also Hua Lo-keng, "Wo-kuo ku-tai shu-hsüeh ch'eng-chiu chih i-p'ieh" (A brief survey of the development of our country's ancient mathematics), *Wen-wu* (1978) 1:47.

/26/   Many variants of this pattern of course existed in the actual layout of towns. See Sen-dou Chang. "The Morphology of Walled Capitals," in G. William Skinner, ed., *The City in Late Imperial China* (Stanford: Stanford University Press, 1977), pp. 75–100, fig. 5. See also Sen-dou Chang, "On the Morphology of Ancient Chinese Walled Cities," *Ekistics* 31, no. 182 (January 1971): 91–98.

/27/   In the ancient Japanese capital of Nara, built in imitation of the T'ang capital at Ch'ang-an, the royal palace was built on a raised terrace against the northern limit of the city (there was no wall); the terrace was called *Hokkyoku-dai*, "north pole terrace."

/28/   From Arthur F. Wright, "The Cosmology of the Chinese City," in Skinner, pp. 33–73, Map 1 (after *K'ao-ku* 1963), no. 11, plate 1).

/29/   I am indebted to Dr. Robin Yates and Prof. Duncan Kim for this information.

/30/   Paul Wheatley, *The Pivot of the Four Quarters* (Chicago: Aldine Publishing Co., 1971), pp. 411–76. Wheatley emphasizes (p. 481) that cities and towns in India and elsewhere in Asia also functioned as mandalas. For a discussion of the pan-Eurasian Grand Origin Myth and its Chinese variants—myths embodying the cosmic moral order that underlay the symbolism of ancient urban planning—see John S. Major, "Myth, Cosmology, and the Origins of Chinese Science," *Journal of Chinese Philosophy* 5:1–20 (1978).

/31/   Cheng Te-kun, "*Yin-yang Wu-hsing* and Han Art," *HJAS* 20 (1957): 162–86.

/32/   Re-drawn after Hsiung Chuan-hsin (see n. 34 above), fig. 1.

/33/   Michael Loewe's *Ways to Paradise* (London: Allen and Unwin, 1980) contains a detailed study of TLV mirrors, including a survey of previous studies and many excellent plates; as it was published just as this paper was going to the press, its findings could not be incorporated here, but it will be indispensable to all future studies of the subject. Among the more significant earlier studies are Nakayama Heijiro, "Koshiki Shina kyokan enkaku" (On ancient Chinese mirrors), *Kokogaku zasshi* 9 (1919); Cheng Te-kun, "*Yin-yang Wu-hsing* and Han Art*,*" *HJAS* 20:162–86 (1957); Anneliese Bulling, *The Decoration of Mirrors of the Han Period* (*Artibus Asiae*, Supplementum XX; Ascona, Switzerland, 1960) (but see also the highly critical review by Cammann of this work in *JAOS* 81, no. 3 [1961] 3:331–36); Schuyler Cammann, "Types of Symbols in Chinese Art," in A. F. Wright, ed., *Studies in Chinese Thought* (Chicago: University of Chicago Press, 1953) 195–231; Schuyler Cammann, "The TLV Pattern on the Cosmic Mirrors of the Han Dynasty," *JAOS* 68 (1948): 159–67; L. S. Yang, "A Note on the So-called TLV-Mirrors and the Game *Liu-po*," *HJAS* 15 (1962): 124–39; W. P. Yetts, *The Cull Chinese Bronzes* (London: The Courtauld Institute, 1939), pp. 116ff.: Needham, *SCC* 3:302–8. A study of TLV mirrors the inscriptions of which relate specifically to Mt. T'ai (which thus is treated as an *axis mundi* equivalent to K'un-lun) is found in E. Chavannes, *Le T'ai Chan* (Paris, 1910; reprinted, Taipei, 1970) 424–26. My analysis of the TLV mirror as a cosmological mandala was completed before I was able to consult Cammann's article in *JAOS* mentioned above; I was pleased to find my analysis confirmed by Cammann's study, which closely parallels the approach presented here and reaches identical conclusions on many points.

/34/   *SCC* 3:302–8.

/35/   L. S. Yang in *HJAS* 9 and *HJAS* 15 (see n. 32 above); Needham, *SCC* 3:302–8 and 4.1:327 and fig. 349; Hsiung Chuan-hsin, "T'an Ma-wang-tui san-hao Hsi-Han-mu ch'u-t'u ti liu-po" (A discussion of the *liu-po* [set] excavated at the Western Han tomb at Ma-wang-tui 3), *Wen-wu* 1979.4:35–39.

/36/   Before the recent archaeological discovery of well-preserved diviner's boards and *lui-po* boards and game sets, scholars were unable to distinguish clearly between the two types of board. See Yen Tun-chieh, "kukan-yu Hsi-Han ch'u-ch'i ti shih-p'an ho chan-pan" ("Notes on the Diviner's Boards of Early Western Han") *k'ao-ku* 1978.5:334–37, and Uen Tun-chieh, "Hsi-Han Hou mu ch'u-t'u ti chan-p'an ho t'ien-wen i-ch'i" ("On the Western Han Diviner's Board Excavated from the Tomb of the Marquis of Ju-yin, and Astronomical Instruments") *k'ao-ku* 1978.5:338–43, and Donald Harper, "The Han Cosmic Board (*Shih* 栻)," *Early China* 4 (1978–79) 1–10. These articles supersede Needham's remarks on diviner's boards in *SCC* 3:302–8, 4.1:261–69, 315. See also the further remarks on the *shih* by Donald Harper and Christopher Cullen in *Early China* 5 (1980–81) and 6 (1981–82).

/37/   From the Cull Collection; see Yetts, pp. 116ff.; Needham, *SCC* 3:Fig. 126; cf. a nearly identical mirror in Bernhard Karlgren, "Han and Huai," *BMFEA* 13 (1941): Pl. 79, L.1.

/38/   The inscription consists of seven rhymed seven-character lines plus a final couplet of four-character lines. Compare the translation of Yetts quoted in Needham, *SCC* 3: fig. 126; and also Karlgren's translation of the closely similar inscription on mirror no. L.1 in "Han and Huai," p. 113.

/39/   Karlgren, "Han and Huai," Pl. 80, L.3.

/40/ Ibid., p. 114.

/41/ Cammann, "The TLV Pattern on the Cosmic Mirrors of the Han Dynasty," pp. 166–67.

/42/ Ibid., fig. 3f, redrawn.

/43/ HNT 4:1ab, unemended.

/44/ Ssu-pu pei-yao edition; reprinted, Taipei, 1966.

/45/ Shan 山 is probably a mistake for t'u 土 . R. Wilhelm, in his translation of Lscc, p. 157, emends shang to t'u and translates this phrase "die Erde hat neun Bezirke."

/46/ Eduard Erkes, "Das Weltbild des Huai-nan-tzu." Ostasiatische Zeitschrift 5:27–80 (1916–17): 33. Richard Wilhelm, Frühling and Herbst des Lü Bu We (Jena: E. Diederichs, 1938), p. 157.

/47/ Lending weight to this translation is Morohashi, Dai Kanwa jiten 3:2382 (character no. 4867, definition 4), t'u = kuo

/48/ Ku Chieh-kang, "Han-tai i-ch'ien Chung-kuo-jen ti shih-chieh kuannien yü yü-wai chiao-t'ung ti ku-shih" (The world-view and foreign relations of the pre-Han Chinese), Yü-kung pan-yüeh k'an, vol. 5, nos. 3 & 4 (combined issue) (April 1936) 97–120. See especially pp. 102ff.

/49/ Ibid., p. 103.

/50/ For example, points of similarity can be found between the division of earth into nine continents, and their further division into subcontinents and provinces, and the division of heaven into five "palaces" (kung 宮), nine "fields" (yeh 野), and 9,999 "points" (yü 隅). See HNT 3:4a, 5b, etc. In Lscc 13:1ab the correspondence between the nine "fields" of heaven and the chiu chou is made explicit.

THE CONTRIBUTORS

JAMES A. HART received his Ph.D. in Chinese Language and Literature from the University of Washington. His articles and reviews have appeared in *Monumenta Serica* and *Philosophy of East and West*, and he is currently in the Law School of the University of Texas in Austin.

JEFFREY A. HOWARD is completing a dissertation on the Sung Historical critic Cheng Ch'iao (1104–62) in the Department of History at the University of Washington.

DAVID N. KEIGHTLEY is the author of *Sources of Shang History*, and an editor of the journal *Early China*. He is Professor of History at the University of California at Berkeley.

JOHN LOUTON recently completed a dissertation on the *Lü-shih Ch'un-ch'iu* in the Department of Asian Languages and Literatures at the University of Washington, and is currently a cell technologist in Seattle, Washington.

JOHN MAJOR is Associate Professor of History at Dartmouth. His articles and reviews have appeared in journals such as *Philosophy East and West*, the *Journal of Chinese Philosophy*, and *Early China*.

HENRY ROSEMONT, JR., has published Leibniz's *Discourse on the Natural Theology of the Chinese* (with D. J. Cook), and *Work, Technology and Education* (with W. Feinberg). He is Professor of Philosophy at St. Mary's College of Maryland.

VITALY A. RUBIN, formerly a member of the Academy of Sciences in the U.S.S.R., published *Individual and State in Ancient China*. He was Associate Professor of Chinese Studies at the Hebrew University of Jerusalem before his untimely death in 1981.

GERALD SWANSON has published articles and reviews in *Monumenta Serica* and *Philosophy of East and West*, and wrote the Introduction to the English translation of I. Shchutskii's *Researches on the I Ching*. He is currently a member of the Department of Religious Studies at the University of Montana.

INDEX